Social Engineering

Social Engineering

How Crowdmasters, Phreaks, Hackers, and Trolls Created a New Form of Manipulative Communication

Robert W. Gehl and Sean T. Lawson

The MIT Press
Cambridge, Massachusetts | London, England

© 2022 Robert W. Gehl and Sean T. Lawson

All rights reserved. No part of this book may be reproduced in any form by any electronic or mechanical means (including photocopying, recording, or information storage and retrieval) without permission in writing from the publisher.

The MIT Press would like to thank the anonymous peer reviewers who provided comments on drafts of this book. The generous work of academic experts is essential for establishing the authority and quality of our publications. We acknowledge with gratitude the contributions of these otherwise uncredited readers.

This book was set in ITC Stone Serif Std and ITC Stone Sans Std by New Best-set Typesetters Ltd. Printed and bound in the United States of America.

Library of Congress Cataloging-in-Publication Data

Names: Gehl, Robert W., author. | Lawson, Sean T., 1977–author.
Title: Social engineering : how crowdmasters, phreaks, hackers, and trolls created a new form of manipulative communication / Robert W. Gehl and Sean T. Lawson.
Description: Cambridge : The MIT Press, 2022. | Includes bibliographical references and index.
Identifiers: LCCN 2021016750 | ISBN 9780262543453 (paperback)
Subjects: LCSH: Social media—Security measures. | Computer networks—Security measures. | Internet fraud. | Social engineering.
Classification: LCC HM742 .G45 2022 | DDC 364.16/3—dc23
LC record available at https://lccn.loc.gov/2021016750

10 9 8 7 6 5 4 3 2 1

This book is dedicated to Teddy and Everett.

Contents

Acknowledgments   ix

Introduction: The Emergence of Masspersonal Social Engineering   1

I   Engineering the Social   25

1   Crowdmasters: The Rise and Fall of Mass Social Engineering, 1920–1976   27
2   Phreaks and Hackers: The Rise of Interpersonal Social Engineering, 1976–Present   49

II The Social Engineering Process   67

3   Trashing: From Dumpster Diving to Data Dumps   69
4   Pretexting: Recognizing the Mitnick Mythology   89
5   Bullshitting: Deception, Friendliness, and Accuracy   115
6   Penetrating: The Desire to Control Media and Minds   139

III Masspersonal Social Engineering   163

7   Contemporary Masspersonal Social Engineering   165

8 **Conclusion: Ameliorating Masspersonal Social Engineering** 199

Notes 227
Bibliography 279
Index 319

Acknowledgments

The authors of any book are always deeply indebted to all of those folks along the way who provided advice, encouragement, support, and feedback. We are no exception. This project would not have been possible without the generosity of colleagues, friends, and family. We would especially like to thank Margot Opdycke Lamme, Professor Emerita at the University of Alabama, for providing a copy of Doris Fleischman's 1935 speech. Biella Coleman of McGill University offered feedback and encouragement early in the process, as did Walter Scheier at Notre Dame. Hector Postigo provided feedback that was critical in helping us decide just what to call the phenomenon we discuss in the pages that follow: *masspersonal social engineering*. Guobin Yang helped to shape our thinking by generously sharing with us his chapter, "Communication as Translation." Finally, both Emma Briant and Cory Wimberly shared their excellent books on propaganda with us.

We are also thankful to the audience and organizers of several events at which we were honored to present early drafts of this work. These included the 2020 Hackers on Planet Earth (HOPE) conference, the Department of Defense Strategic Multilayer Assessment, and the Critical Genealogies Workshop. At the latter, we especially thank Bonnie Sheehey, Ashley Gorham, Verena Erlenbusch-Anderson, Cory Wimberly, and Colin Koopman for their extensive comments.

Finally, of course, we are grateful for the patience and assistance provided by our editor at the MIT Press, Gita Manaktala, as well as the anonymous reviewers whose comments helped shape and improve this book. Nonetheless, of course, any errors or omissions are all our own.

I would like to thank the Louisiana Board of Regents Endowed Chairs for Eminent Scholars program, which provided funding for this project. The University of Calgary Communication, Media, and Film Department and the Louisiana Tech Communication faculty provided valuable comments on an early version during my Fulbright fellowship. I also had a great conversation with Claire Evans about Susy Thunder and phone phreaking.

I am, as always, in deep debt to my friends, old (Ry, Mony, Brian, and Dan) and new. A new and deep friendship has been made during the writing of this book: Sean, thank you (and Cynthia!) for not only putting up with me but also putting me up during a time of disasters.

Love from my family keeps me going even in the midst of bleak times. My mom, my little brother, the Houf family. And I cannot think of anyone I'd rather weather the storm with than Captain TJ and Navigator TJ.

Robert W. Gehl
June 2021

I would like to thank my close friends and family for their patience and support during the process of writing this book, most of which took place during the challenge of a global pandemic. First and foremost, my wife, Cynthia, whose advice and encouragement keeps me going. Rob, my coauthor and friend: thank you for inviting me to be a part of this project and for your patience along the way. I'm ever grateful for the chance to learn and grow as a writer through our collaboration. Finally, I, too, must thank Captain and Navigator TJ, who helped keep us all well fed and sane during some tough times early in the project.

Sean Lawson
June 2021

Introduction: The Emergence of Masspersonal Social Engineering

The United States is awash in a disorienting and sometimes deadly digital media environment. People are sharing—sometimes purposely, other times without knowing any better—manipulative information about everything from election results to the effectiveness of medical treatments. Domestic political leaders seek to ride the resulting paranoia and confusion to ever greater power, while foreign governments gleefully stoke divisions and discontent.

Manipulative communication found a home in corporate social media, especially Facebook, Facebook's child company Instagram, and Twitter. These systems were designed to amplify attention-getting messages, whether those messages are cute cat videos or the latest QAnon "drops" of conspiracy theorizing. Facebook has especially proven to be a willing vehicle for manipulative communication, using its vast data on our likes and preferences to route to us the information that satisfies our desire for affirmation. No matter your political or social tastes, Facebook will deliver a meme that confirms your views—even if the meme is bullshit. And if Facebook's own algorithms don't deliver a message to its intended audience, the creators of the message can simply pay a small fee to microtarget their ideas to audiences likely to agree with them.

Such manipulative communication erupted in a paroxysm of violence on January 6, 2021. That day, a riotous mob of Donald Trump supporters occupied the US Capitol building. The mob was spurred on by a blatant misinformation campaign, run by Trump and several of his allies (notably Senators Josh Hawley and Ted Cruz), all of whom peddled bullshit about rampant election fraud in the 2020 US presidential campaign.[1] This Capitol insurrection was meant to keep Donald Trump in power. Five people died—four rioters and one police officer—but reports indicate that the death toll may have been higher if the rioters had been able to find their perceived enemies, including Vice President Mike Pence, House Speaker Nancy Pelosi, and Representative Alexandria Ocasio-Cortez.[2]

The Capitol insurrection is a cautionary example of the dangers of manipulative political communication. But, as we will argue, the sorts of manipulative communication that fueled the events of January 6, 2021 are nothing new. Consider, for example, the 2016 US presidential election, which pitted the Democratic candidate Hillary Clinton against the reality-television-star-turned-Republican Donald J. Trump. There are now two well-documented types of manipulation campaigns that took place in the run-up to that election.

First, there were attempts by a foreign government, Russia, to undermine confidence in the election, support Trump at the expense of Clinton, and stoke racial and political divisions among Americans. Thanks to several investigations, the campaign is well documented, its complexity revealed. The operation involved the hacking of email accounts associated with the Democratic National Committee and the Hillary Clinton campaign, theft of data, and the leaking of those data to third parties like WikiLeaks. Russian state media outlets *RT* and *Sputnik* helped to promote the leaks.[3] The employees of the Saint Petersburg–based Internet Research Agency helped amplify these messages by manipulating Facebook, Instagram, and Twitter through human-controlled and bot-controlled accounts.[4] They piggybacked on 4chan's "Great Meme

War," amplifying pro-Trump and anti-Clinton memes.[5] The Internet Research Agency operatives also posed as real Americans, appropriating the online conversations of Black Lives Matter activists and conservative pundits alike.[6] All of this effort was in support of the unlikely candidacy of Donald Trump, who would of course go on to win the election.

But foreign actors weren't the only ones taking to corporate social media to manipulate democratic deliberation during the 2016 election. The other well-known manipulative communication campaign was based in the West. Thanks to the glut of data available—and sometimes unethically obtainable—from Facebook, the strategic communication firm Cambridge Analytica claimed to be able to target Americans along a range of "psychographic" vectors and subtly shape their opinions via social media. Led by its then-CEO, Alexander Nix, a man *Wired* magazine proclaimed a "genius," Cambridge Analytica promised to exploit personality traits of targeted voters.[7] Trump's presidential campaign contracted with them to create and target advertisements to sway voters away from Clinton and towards the reality star.

We need to offer a caveat, though. A debate is currently raging about the effectiveness of either the Russian effort to sway the election towards Trump or Cambridge Analytica's role in shaping voting behaviors.[8] In any given election, there are countless factors at play—we cannot reduce social change to a single cause. For example, we should not forget the widespread and ongoing efforts at voter suppression, or the simple fact that Clinton chose not to campaign in the key swing state of Wisconsin during the general election. Nonetheless, the election of 2016 alerts us to the emergence of disturbing new practices of manipulative communication. And the danger of manipulative communication is all the more apparent in the wake of the 2021 Capitol insurrection.

Starting in the mid-2010s, we saw new, emerging forms of manipulation: email account hack-and-leak operations, military investment in social media manipulation capabilities, messages appearing to come from neighbors but actually made in Russia, memes picked up from obscure parts of the internet and plastered over Facebook and Instagram, experimental posts meant to elicit and exploit particular emotional responses. Many of these manipulative communications were driven by data collection efforts and featured operatives hiding their true identities behind fake accounts and front organizations. The manipulators interpersonally engaged with their targets and then claimed victory when they saw their manipulative messages amplified across mass media.

There has been a clear convergence of these tactics of manipulation into something new: *masspersonal social engineering*. We define masspersonal social engineering as

> an emerging form of manipulative communication enabled by the unique affordances of the internet and social media platforms. It brings together the respective tools and techniques of hackers and propagandists, interpersonal and mass communication, in an attempt to shape the perceptions and actions of audiences. To manipulate, masspersonal social engineers gather data on their targets; create fake personas to share messages; mix deception, accuracy, and friendliness as they engage with targets; and penetrate communication systems. Manipulation, in this case, can involve a range of goals, which might include attempts to change actions and beliefs. But it could also include discouraging action (e.g., voting) and amplifying or intensifying preexisting beliefs (e.g., racism, sexism, or other social divisions) when doing so is in the perceived interests of the masspersonal social engineers or their clients.

Masspersonal social engineering has only intensified since the 2016 election. Despite several investigations into the Russian meddling, and despite whistleblowing, investigative reporting, and ultimately the dissolution of Cambridge Analytica, manipulation of democratic deliberation by both foreign governments and domestic political consultancies continues apace. The Russian government has not stopped seeking to use digital media to engineer chaos within the American electorate. Russian agencies continued to apply their techniques in both the 2018 midterm elections and the 2020 presidential election.[9] And while Cambridge Analytica is no more, its techniques have been taken up by firms such as Phunware and Rally Forge, predominantly in service to conservative causes and politicians—although left organizations have also deployed these tactics.[10] Researchers have even identified a nascent market for underground "disinformation as a service," where vendors sell services to manipulate digital media to either puff up a cause or tear down and disparage a rival.[11]

In sum, starting with the 2016 election and continuing to today, a new form of manipulative communication, a new form of social engineering, has increasingly come into focus.

Why "Social Engineering" Is a Good Label

We're not the only ones concerned about media manipulation. Many people are trying to understand these new forms of misinformation and disinformation, and the vocabulary has yet to be decided upon.[12]

One of the more common phrases being used is "fake news," which has the appeal of cutting through the fog and labeling Russian-backed posts, amplified memes, and psychographic ads for what they are: disingenuous and misleading. Similarly, as the title of a recent academic book puts it, maybe what we're seeing are "lie machines," systems developed to pump out falsehoods that fool us.[13]

But something about the "fake" or "lie" label seems to fall short, failing to grasp the complexity of manipulative communication. "Fake news," as many others have pointed out, reduces everything to a stark true/false binary. The same is true of the concept of "lie machines"—one gets the vision of a vast network of bots and trolls churning out Augustinian lies. If manipulative communication was really an exercise in telling pure lies, it should have been an easy matter to debunk lies and hold up truths. But despite all the fact-checkers pointing out falsehoods, manipulation continues apace. Moreover, the true/false dichotomy doesn't quite map onto a lot of the practices of manipulative communication. Is a message meant to elicit your sense of sympathy a lie? Is a funny meme a falsehood? Is a sponsored post that your friend endorses untrue?

"Propaganda" seems to be an obvious term to use, even though its meaning has been muddied over the years.[14] A recent book, *Computational Propaganda*, purports to update the term for the digital age, addressing the use of algorithms and automation in modern manipulative communication.[15] However, propaganda—which is now more often called "public relations"—only invokes the mass side of communication.[16] Indeed, another recent study of propaganda, *Networked Propaganda*, actually *excludes* interpersonal manipulation as distinct from propaganda, thus excluding many of the manipulative, one-on-one interactions now possible in social media.[17]

Closer to the mark might be "hacking." After the 2016 election and in the run-up to the 2020 US presidential election, many commentators used phrases such as "the Russians are seeking to hack the election." But this term is also inadequate. "Hacking" invokes the image of the hoodie-clad loner, sitting above a glowing keyboard and furiously hacking . . . what, exactly? The hackability of digital voting machines has been a longstanding concern in the US. But what happened in 2016 and again in 2020 wasn't about hacking voting machines—not that people didn't try.[18] Instead, commentators modify "hacking" with "cognitive," as in "hacking our minds."[19] This gets a bit closer to what happens when manipulative masspersonal social engineering occurs. Cambridge Analytica, in particular, claimed that it had thousands of data points on every American voter, allowing it to find the subtle mental levers to press in order to shape opinions. And while Cambridge Analytica may have exaggerated its abilities, its brand of microtargeting is now a fixture of contemporary political communication.[20]

But "hacking" seems too instrumental. Like "lie machines," it's stark: if, for example, a malicious hacker gains root access to a computer, they get total control, which does not quite map onto the subtle manipulations attempted during the 2016 election and after. No one gets root access to someone else's mind. Even with thousands of data points, Cambridge Analytica could not control anyone's mind. Communication—the messy process of mutually building reality—cannot be reduced to a computational metaphor.

"Cyberwar" is another term, most notably used as the title for communication scholar Kathleen Hall Jamieson's latest book. In *Cyberwar: How Russian Hackers and Trolls Helped Elect a President*, Jamieson argues that because the Internet Research Agency efforts were sponsored by a state, Russia, and directed at another state, the US, and intended to interfere in a political process, we must therefore consider the effort an act of war and respond accordingly.[21] Thus, unlike "hacking an election," which invokes individual hackers manipulating targeted machines or people, Jamieson's conception acknowledges the societal, political scale of the continued effort by Russians to interfere in democratic deliberation.

Though "cyberwar" might be a tempting label for Russian state-sponsored operations, we argue that it is neither a good fit for Russia's actions nor for the actions of firms like Cambridge Analytica. The claim that Russian cyber operations in 2016 were a type of "war" is not well rooted in existing law or scholarship on cyber conflict.[22] Even if the "cyberwar" label applies to what Russia attempted during the 2016, 2018, and 2020 elections, what do we call similar efforts by Cambridge Analytica or other domestic actors?[23] Do we call the consulting firm Cambridge Analytica—which was hired, after all, by the Trump presidential campaign—a state actor? Is Phunware, an Austin, Texas–based firm which worked with the 2020 Trump campaign, waging war?[24] Commentators have argued that firms such as Cambridge Analytica "weaponize" data.[25] But we cannot declare war based on a war metaphor. Calling every malicious act online "cyberwar" only invites further militarization of the internet.[26]

Most of the concepts used to wrestle with manipulative communication fall short. But there is one term that's been used, albeit much less frequently than "fake news," "lie machines," "propaganda," "cognitive hacking," or "cyberwar," that we find to be complex and flexible enough to describe the new form of manipulative communication that emerged in 2016 and continues to run rampant in digital media today.

That term is "social engineering."

The term alone, with no definition in mind, is evocative: engineering the social, shaping social interactions through systematized techniques. The term hints at a dream: what if there are techniques that could allow us to control human nature, just as we could control the natural world through engineering? Indeed, the term does refer to those practices, as we will show. But "social engineering" is not just evocative; it has a deep history, one that extends past the realms of security hackers and political communication propagandists, beyond the targeting of individuals for con artistry and the desire to shape the consciousness of a nation, and across the blurry lines between fact and falsehood, friendliness and maliciousness, earnest do-goodism and cynical opportunism, and simplistic and complex understandings of communication.

On Social Engineering

First, let's review some commentators who have connected the new manipulative communication to the concept of social engineering.

One connection between social engineering and the Russian election interference operations was made by security consultant

Kevin Mitnick. As the self-proclaimed "World's Most Famous Hacker," Mitnick will figure heavily in the pages of our book. Mitnick is notable as the hacker most associated with a specific practice of using con artistry to gain access to restricted computer systems. Rather than attack a network at a technical level—say, by trying to break through network encryption or brute force a password—Mitnick would target *humans*: he would call computer operators up on the phone and simply ask for the password.

He called this practice "social engineering," a term that hackers started using in the mid-1970s. As security consultant Sharon Conheady defines it, hacker social engineering "involves convincing people to perform actions they would not normally do."[27] These actions include giving out passwords, letting people roam around restricted areas of a corporate campus, providing access to restricted information or financial capacities, or visiting a malicious website. For hacker social engineers, the target of social engineering is the individual: a secretary, an IT employee, a CEO, a political campaign staffer.

These days, Mitnick's style of social engineering happens predominantly over email. We call it phishing. The basic technique is to try to convince people to click links they normally should not, links that might either load malware onto the person's computer or lead to spoofed websites meant to harvest passwords. You've almost certainly gotten such an email.

Phishing was central to the 2016 election, because the Russians used phishing to compromise the computer of John Podesta, then the chair of Hillary Clinton's election campaign. According to AP News, after a series of unsuccessful phishing emails were sent to former Clinton campaign staffers, the Russians homed in on Podesta. On March 19, 2016, "a malicious link was generated for Podesta at 11:28 a.m. Moscow time. Documents subsequently published by WikiLeaks show that the rogue email arrived in his inbox six minutes later. The link was clicked twice."[28]

Reflecting on this phishing attack, Mitnick notes that

> what was interesting is that the methodology the Russians used to hack the DNC is really no different from what civilians, whether crooks or people like us doing security testing—we use the same method of spearphishing. It's social engineering.[29]

But the Podesta phishing attack was not just meant to get Podesta to pay a ransom or to con him out of money, as most phishing is meant to do. The emails were leaked to Wikileaks, with the goal being their public dissemination. Here, phishing was used to expose the inner workings of the Clinton campaign, fomenting conspiracy theories and further tarnishing Clinton's reputation during the election.[30]

Here we see "hacking an election," but we want to be more precise: it was a phishing hack of a Clinton staffer, not the election as a whole. Given what resulted—the Clinton campaign's internal deliberations being posted on WikiLeaks—this specific attack certainly had massive ramifications for the election. The emails became fodder not just for media outlets but also for pro-Trump meme warriors, whose memes would be amplified by Russians using social media. So, we may say that the 2016 election was socially engineered in the hacker sense of an interpersonal con job, the convincing of someone to do something they should not, resulting in a computer security breach.

But ultimately, it was one person targeted with one hacker technique, and arguably such social engineering was only one tactic used in support of a larger strategy that played out on a national scale, a strategy that included propaganda campaigns, political communication, and social media trolling. The question then becomes: can the term "social engineering" capture this scale, or is it inadequate?

In short, it can. Consider another use of the label "social engineering" to discuss the Russian operation, this time by security researchers. As Thomas Rid, a leading security scholar, put it, the

Russian operation that started in 2016 and continues to this day is "political engineering, social engineering on a strategic level."[31] As for political communication firms, such as Cambridge Analytica, several commentators referred to their communication practices as social engineering. One security researcher used the label to refer to "companies like Cambridge Analytica and the propaganda machine of [Trump advisor] Steve Bannon" as engaging in "social inception" via social engineering.[32] Others called it "extreme social engineering"—the use of large datasets allowing for "illicit deception campaigns on a large scale."[33] The most explicit condemnation comes from Cambridge Analytica whistleblower Christopher Wylie's book *Mindf*ck*, which likened the firm's attempted manipulation of society to engineering a railroad.[34]

These critiques of Russia's Internet Research Agency and Cambridge Analytica draw on the hacker sense of social engineering (e.g., phishing, conning people out of passwords), but with a twist. Instead of targeting individuals, these operations worked on a massive scale, claiming to target populations. Indeed, one team of security researchers makes this point explicitly: "while [hacker social engineering] was historically practiced face-to-face, over the phone, or through printed writing, social engineering can now occur on societal scales through social media and other internet platforms."[35]

While they may be drawing largely on the hacker sense of the term "social engineering," in drawing attention to the societal scale of these efforts, these commentators invoke a different meaning of "social engineering," an older one that dates back to the late nineteenth century and reached its apogee in the 1920s to 1950s. This meaning of social engineering is associated with mass media, propaganda, and public relations. It is best illustrated by the husband-and-wife team Edward Bernays and Doris Fleischman, who referred to their early twentieth-century work on propaganda as the "engineering of consent." Consent engineers like Bernays and Fleishman laid the groundwork for the contemporary field of public relations—previously called "propaganda."[36] In this version of social engineering, rather than concentrating on getting an individual to take an action—like giving out a password, as a computer hacker might—the goal is to master crowds of people, influencing them to buy more products, respect their corporate betters, or support a war effort.

And just as it used the hacker social engineering technique of phishing, the Russian operation has included the use of mass propaganda. The US Office of the Director National Intelligence (ODNI) reported in 2017 that "Russia's state-run propaganda machine—comprised of its domestic media apparatus, outlets targeting global audiences such as *RT* and *Sputnik*, and a network of quasi-government trolls—contributed to the influence campaign by serving as a platform for Kremlin messaging to Russian and international audiences."[37]

This form of social engineering is in many ways classic propaganda, a form early twentieth-century communication theorists would have easily recognized as "the management of collective attitudes by the manipulation of significant symbols."[38] *RT* and *Sputnik*, the DNI report assessed, amplified existing domestic controversies in the United States, such as debates over police power, concerns over the use of fracking, and protests over bank bailouts, in order to sow further discontent.[39] The overall goal was to increase acceptance of Russian geopolitical policies by presenting Russia as a sane, measured world leader and the United States as a hypocritical, dissent-laden imperialist power.[40] In this sense, the Russian effort was an attempt at social engineering in the older, "engineering of consent" sense of Bernays and Fleischman.[41]

Such attempts weren't limited to the Russian operation. As Cambridge Analytica grew in stature in the early 2010s, it promised the political campaigns it took on as clients that it could shape national elections in their favor. As Cambridge Analytica whistleblower Brittany Kaiser recalls in her memoir, the firm "had amassed an arsenal of data on the American public of unprecedented size and scope, the largest, as far as [they] knew, anyone had every assembled."[42] This massive database was deployed by Cambridge Analytica on behalf of Trump's 2016 presidential campaign. Like Bernays and Fleischman before it, Cambridge Analytica saw its job as engineering the consent of voters, swaying people to make decisions, as Kaiser puts it, "against their usual judgment, and to change their habitual behavior" and affect the course of history.[43]

Here, then, we have a term, "social engineering," that at the very least bears two meanings, both of which illuminate new manipulative communication practices. The hacker meaning invokes personalized con artistry, the sort of thing the Russians did to the Clinton campaign's John Podesta. The political meaning invokes large-scale, mass societal manipulation via media, the sort of thing attempted by the Russian Internet Research Agency and political communication firms.

But as we will show, social engineering does not simply have a split meaning. Instead, it can encompass both interpersonal and mass vectors. We could liken emerging practices of manipulative messaging to the large-scale social engineering practice of propaganda, but that does not quite capture the full sense of what happened. Mass social engineering—the art of propagandists and public relations practitioners—targeted large groups and populations. Taking up the social sciences developed at the turn of the twentieth century, this form had grand societal ambitions and held crowds or masses as the central object to manipulate. So, this meaning certainly captures the scale of the new form of manipulative communication that emerged since 2016.

But much of what makes that new form unique was the attempted *targeting of individuals* via social media with the ambition of having societal-scale effects.

Thus, the other meaning of social engineering, the interpersonal hacker meaning, often appears to come closer to the mark, since that approach is predominantly directed at individual targets. Returning to the Director of National Intelligence quote cited above, note a modern twist on the older propaganda theme: the use of "a network of quasi-government trolls."[44] What were these trolls up to? The Internet Research Agency trolls were personally controlling social media accounts that "were spontaneous and responsive, engaging with real users (famous influencers and media as well as regular people), participating in real-time conversations, creating polls, and playing hashtag games."[45] This is not mass messaging, but highly personalized, more like the form of social engineering developed by hackers.

Moreover, the sales pitch that firms like Cambridge Analytica bring to clients is decidedly not about mastering crowds and masses. Cambridge Analytica's massive database on American voters, for example, was not geared towards what they dismissed as "blanket advertising . . . messaging intended for a broad audience and sent out in a giant, homogeneous blast."[46] Instead, Cambridge Analytica wooed clients with the promise of the precision of "microtargeting," taking their data and using "its own algorithm to scan them, identifying likely political persuasions and personality traits [of voters]. They could then decide who to target and craft their messages that was likely to appeal to them for those individuals."[47] In this sense, the firm contrasted itself with the mass propagandists of years past, claiming to fuse the science of psychographics with the profiling and targeting capabilities of contemporary social media. Microtargeting has not gone away with Cambridge Analytica's demise, either: the Trump 2020 campaign also sought to microtarget voters using Facebook's advertising tools.[48] Again, like the personally controlled social media accounts of the Russians, this approach sounds far more like interpersonal hacker social engineering than the mass messaging of the consent engineers.

Overall, the Russian election interference operations and the political communication microtargeting campaigns reveal that the lines between a mass form of social engineering and the more interpersonal hacker social engineering have become blurry. If we consider the recent attempts to interfere with or manipulate the American electorate as social engineering, and if the term "social engineering" invokes both mass and interpersonal communicative practices, the older divisions between "mass" and "interpersonal" may no longer be useful. We thus turn to a newer, convergent conception of the communication process, *masspersonal communication*, and then our ultimate conceptual goal, *masspersonal social engineering*.

On Masspersonal Communication

Communication studies has traditionally observed a stark boundary. On one side, there's interpersonal communication: people talking to each other, either face to face, over the phone, or through digital media such as email or texting. The other side is mass communication: a one-to-many model. The advent of technologies like the printing press, and later cinema, radio, and television, allowed for one-to-many, impersonal, and asynchronous communication on a mass scale. Through the twentieth century, then, communication theorists distinguished between these two dominant models of communication based on the different "channels" by which they occurred: direct and synchronous, or mass mediated. To this day, many schools and departments of communication in the United States focus on either mass communication (typically specializing in journalism) or interpersonal communication.[49]

But this distinction, along with many others, began to blur with the advent of computer-mediated communication technologies in the latter half of the twentieth century. That process was accelerated in the 1990s and early 2000s with the growing popularization of the internet. These same technological changes were linked to the emergence of postmodern and postindustrial societies in Western nations during this same period. By the 1990s, popular commentators were extolling the virtues of the information revolution and its supposedly "new economy" that allowed for just-in-time production and mass customization. In the new economy, information and communication technologies would allow for the mass production of goods personalized to the buyer's specifications.[50]

In light of these changes, scholars began to question the traditional distinction between mass and interpersonal communication.[51] They began to note that the internet not only allowed individuals heretofore lacking access to the technologies of mass broadcast to engage in one-to-many communications of their own but also to engage virtually and remotely in communications that would traditionally have been considered interpersonal had they occurred face-to-face.

Though it was becoming increasingly clear that the mass/interpersonal distinction was dissolving, it remained difficult for scholars to fully account for this change. In 2018, communication theorists Patrick O'Sullivan and Caleb Carr argued that this was because scholars had continued to define the two dominant models of communication primarily based on the "channel" by which such communication occurred—phones are interpersonal, television is mass, for example.[52] Instead of channels, masspersonal communication theorists argue we should focus on the sender's "perception of accessibility, or the perceived number of people who can view a message," and "the receiver's subjective judgments of personalization, or the extent to which a message is perceived as tailored to them personally."[53] While mass communication messages are intended to be widely accessible by a large audience, interpersonal messages are intended to be the opposite. Much hinges on how the message is tailored to the "uniqueness or distinctiveness" of the receiver, often based on an analysis of the receivers' "interests, history, relationship network, and so on."[54]

When we focus on accessibility and personalization of messages rather than channel, we start to see that many messages fall in between mass and interpersonal. A tweet directed at a particular person might be seen by millions, a phone call can be recorded and rebroadcast, a politician's son could make a heartfelt plea for his father's affection during a televised speech. This fluidity between personalized and mass spectrum is, for O'Sullivan and Carr, "masspersonal communication."

Though O'Sullivan and Carr note that examples of such communication can be found throughout history, the poster child for masspersonal communication is, of course, social media. As examples, they offer Twitter mentions, Facebook comments and likes, personalized video, or the more malicious "tailored spam." This tailoring or personalization, they explain, is enabled by tracking and compiling data about people's online activities, which then allows communicators to "narrowcast" their personalized messages to individuals or small groups.[55]

The degree to which the message is personalized also affects how the receiver of the message might respond, leading to interactions or one-way reception of the message.[56] Such personalization can involve targeting an individual or small group with a message that is widely accessible to larger audiences and yet is specifically designed to elicit some kind of response on the part of the receiver. As O'Sullivan and Carr explain,

> ... a personalized interaction is accessible to a large audience, yet is intended as a personalized message in that its content is applicable only to the intended recipient, which may be an individual or select audience (e.g., followers). Moreover, masspersonal communication often facilitates either a private or public response from the (un)intended receiver(s).[57]

Personalization in mass communication, or the scaling up of traditionally interpersonal communication to a larger scale, is the reality of life with digital media. Social media influencers make authentic connections with their fans. Small groups of people making memes for their own consumption see their memes bubble up into popular consciousness.[58] Emails meant for one person leak and become news headlines. Personalized advertising messages in corporate social media increase sales as part of a global advertising campaign. And, of course, the flipside is the personalization of previous mass messaging tactics: marketers want customers to have personal relationships with mass-produced brands. Large campaigns get decomposed into "A/B" testing of messages on increasingly granular audience segments.[59] In sum, the old disciplinary division in the field of communication is no longer useful. Likewise, the division between mass manipulation and interpersonal con artistry is also blurring in the digital age.

Masspersonal Social Engineering

Masspersonal communication includes many communicative practices. Our interest is in the specific communicative practice of social engineering, a practice we will show to be a mixture of information gathering, deception, and truth-indifferent statements, all with the instrumental goal of getting people to take actions the social engineer wants them to take.

We argue that the ongoing Russian operations, Cambridge Analytica, and subsequent, similar attempts at manipulative communication bring into stark relief the fact that the dual-meaning of social engineering—older, mass consent engineering and the newer, individualized hacker con artistry—was not so much a bifurcation as a false dichotomy. Just as the distinction between mass and interpersonal models of communication were never as stark as some had assumed, there has always been an overlap between mass and interpersonal forms of social engineering. This is in spite of the fact that the two meanings developed independently of one another. While the two meanings are separated in time and have distinct histories, the "consent engineering" Bernays and Fleischman proposed in the early twentieth century, with its focus on mass media technologies to persuade targeted publics to adopt specific positions, is not too distinct from Kevin Mitnick's style of interpersonal con artistry intended to get a targeted user to give up a computer password.

We take up US Naval intelligence officer Joseph Hatfield's observation that mass social engineering and interpersonal hacker social engineering are "both conceptually and semantically interrelated."[60] Ultimately, social engineering has converged just as communication has in the direction of what we will call masspersonal social engineering. In this book, we conceptualize this new form of manipulative communication, tracing how older mass social engineering and more recent interpersonal hacker social engineering are converging in today's digital media.

Plan of the Book

In part I, Engineering the Social, we will discuss the older mass social engineering of the early twentieth century. This form of social engineering, which saw its heyday from the 1920s through 1950s, has been well documented by historians.[61] The mass social engineers had a grand vision: that the insights derived by social sciences could be applied to alleviate specific social ills, such as race relations, class relations, or the lack of awareness of new consumer goods. If we had to symbolize mass social engineering with a single communication technology, it would be the newspaper. Its key practitioner is the public relations consultant, a persona dubbed by one early twentieth-century communication critic the "crowd master."[62]

However, as we will show, by the mid-twentieth century, mass social engineering becomes far less valued; it even becomes feared.

The radical individualism emphasized from the 1970s on brought about deep anxieties about societal-scale attempts to ameliorate social problems. This is true of both the political right, which fostered libertarian economic thinking that is fearful of the state, and the political left, which saw large-scale corporate and government practices as a deadening bureaucracy dedicated only to consumerism or war-making.

Another meaning of "social engineering" emerged in the 1970s, right as the older form appeared to die. This form is hacker social engineering. All our historical research indicates that the hackers (originally known as "phone phreaks," as we will discuss) coined "social engineering" as a name for their interpersonal practices of con artistry and bullshitting with no knowledge of the fact that the social reformers of previous generations used that term in a very different way. We argue that the phreaks and hackers would go on to systematize this form of social engineering into what we call interpersonal or hacker social engineering. If we had to symbolize phone phreak/hacker social engineering with a single technology, it would be the telephone. Its practitioner, of course, would be the phreak or hacker.

Since our ultimate goal is to consider the emergence of masspersonal social engineering in 2016 and beyond, our task is to conceptualize what social engineering is. The course we have taken in this book is to pay special attention to the concepts and terms used by the hacker social engineers. We do so for two reasons. First, because in comparison to the scholarship on mass social engineering, there is less scholarship available on hacker social engineering. Our work can make a bigger contribution if we concentrate on the less well-known hacker social engineering. But more importantly, we find that hacker social engineers have developed brutally honest—even enjoyably honest—language for what they do and the concepts they use are sophisticated, intellectually rich, and eye-opening. For one key example, hacker social engineers initially referred to social engineering as "bullshitting." Our impulse might be to reject the term "bullshitting" because it is crude—it doesn't sound academic—but as we will show, philosophers have taken the idea of "bullshit" very seriously, and we will, too. Moreover, armed with the hacker concept of bullshitting, we gain a new lens to reconsider the older, mass social engineers themselves as able bullshitters, and we can consider more recent manipulative communication as rife with bullshit. We can do the same with other fascinating hacker concepts, like trashing and penetrating.

So, this book will be a genealogy of social engineering, rather than a year-by-year history. Our approach is to use the phone phreak and hacker social engineering concepts of trashing, pretexting, bullshitting, and penetrating to better understand both mass social engineering of the early twentieth century and the emergence of masspersonal social engineering in the early twenty-first century. While we are shocked by what we've witnessed these past few years, while we are concerned about this emergent practice of masspersonal social engineering, we do not want to treat the practice as entirely new. We see in hacker social engineering a set of conceptual threads that connects our current moment to the past, including the older mass social engineers. By tracing contemporary manipulative communication to previous forms of social engineering, perhaps the shock wears off and the possibilities for amelioration become clearer.

Thus, part II of the book will be organized around the hacker concepts. In chapter 4, we will start with the peculiar way hackers and phone phreaks gathered intelligence in the 1970s through 1990s: going through trash. We connect the act of diving into a dumpster to contemporary practices of wading through Big Data culled from the internet, and we will also look back to the mass social engineers to see moments when they, too, looked through garbage. We will next explain the concept of the pretext, or the role a hacker social engineer plays during an engagement with a target, in chapter 5. We will find that consent engineers like Edward Bernays, his partner Doris Fleischman, and their contemporary Ivy Lee also relied on pretexts. We follow with the conceptually rich and profoundly compelling practice of bullshitting, which we see as a truth-indifferent mix of deception, accuracy, and sociability, discussed in chapter 6. The best hacker social engineers will proudly claim to be excellent bullshitters. Mass social engineers would never refer to themselves that way, but they bullshitted admirably. Then, in chapter 7, we will turn to penetration, the goal of hacker social engineering, where social engineers get control over the supposedly irrational Others who control systems and networks. We will consider the penetration metaphors offered by both mass and interpersonal social engineers.

Along the way, this section will also show how social engineering articulates with broader cultural dynamics, including the glut of digital data exploding across networks, racial stereotypes, and gender roles.

Finally, part III brings interpersonal social engineering and mass social engineering together into the new practice of masspersonal social engineering. In chapter 8, we return to the themes of this introduction—the emergence of a new form of manipulative communication—but freshly equipped with concepts and practices from hacker social engineers. We consider the 2016 Russian operation and Cambridge Analytica from the perspectives of trashing, pretexting, bullshitting, and penetrating. We focus on those two because they are well documented, but we will also point to more recent examples that fit the mold of masspersonal social engineering. Our goal is to illuminate the ways in which the interpersonal social engineering practices of hackers were scaled up to meet the societal ambitions of mass social engineers. We see this happening on the internet, bringing with it a political milieu of distrust, anxiety, and decay. If we had to symbolize this form of social engineering with a communication technology, it would be social media.

And its practitioners include people like psychographic marketers and Russian operatives seeking to manipulate elections.

We will conclude with recommendations for ways to undermine masspersonal social engineering. Once again, we will use the phreak and hacker concepts of trashing, pretexting, bullshitting, and penetrating as pressure points to find vulnerabilities in masspersonal social engineering. While masspersonal social engineering is a complex manipulative communication practice, we can find ways to undermine it and move towards healthier democratic deliberation.

I
Engineering the Social

1
Crowdmasters: The Rise and Fall of Mass Social Engineering, 1920–1976

> I would like to try to express the type of modern man who . . . is about to prove himself the real ruler of our modern world, the silent master of what the crowds shall think.
>
> —Gerald Stanley Lee on "The Crowd-Man," *Crowds: A Moving-Picture of Democracy*[1]

In a house in Cambridge, Massachusetts, in 1990, a 40-something Marxian critical scholar of consumerism interviewed a wizened man of nearly 100 years, a man who had helped build the very consumer society the scholar was criticizing. The younger man was Stuart Ewen, a professor of film and media studies at Hunter College. The older was Edward Bernays, one of the most important public relations pioneers of the twentieth century.

Although their backgrounds were different, Bernays warmed up to Ewen, seeing him as a member of an "intelligent few" who was "charged with the responsibility of contemplating and influencing the tide of history."[2] Ewen was, after all, a published author and a professor. Bernays saw himself as an intellectual, a theorist of public relations—the field he had helped create in the 1920s.

In his interview with Ewen, Bernays explained his theory of the role of public relations in society. Since Ewen was a member of the "intelligent few," Bernays felt that he could be frank about the field's elitism. Ewen writes:

> As a member of that intellectual elite who guides the destiny of society, the [public relations] "professional," Bernays explained, aims his craft at a general public that is essentially, and unreflectively, reactive. Working behind the scenes, out of public view, the public relations expert is an "applied social scientist," educated to employ an understanding of "sociology, psychology, social psychology, and economics" to influence and direct public attitudes. Throughout our conversation, Bernays conveyed his hallucination of democracy: A highly educated class of opinion-molding tacticians is continuously at work, analyzing the social terrain and adjusting the mental scenery from which the public mind, with its limited intellect, derives its opinions.[3]

Ewen, who had studied the writing of Bernays for years and had authored books on "captains of consciousness" and consent engineering, was probably not surprised to hear this hierarchical vision of society, where "people in power . . . shape the attitudes of the general population."[4]

Today, Bernays's elitism sounds out of date, even dangerously anti-democratic. This is especially so when we are told that our opinions matter, that social movements can use contemporary media to spread their messages, and that we can finally speak truth to power. Even in the 1990s, when Ewen interviewed Bernays, the old man's ideas seemed offensive.

Indeed, Bernays was a product of another time, the early twentieth century, a time when elite experts—especially engineers—were seen as humanity's saviors. Right up to the end of his life (Bernays died in 1995, a few years after Ewen's visit), he held fast to his belief that the masses needed leadership, and that leadership would come from an elite, technocratic few who would shape the masses' reality and thus produce a better society.

These elites used a variety of names for themselves: "public relations professional," "news engineers," "engineers of consent," "crowd-men." We will call them mass social engineers. Their roots lie in the turn of the twentieth century, a period when engineering was held in high esteem, so much so that people believed *society itself* could be engineered just as easily as a bridge or canal could be.

Engineering Society

Writing in 1976, civil engineer and author Samuel C. Florman pined for the "Golden Age of Engineering"—a period he defined as 1850 to 1950.[5] During this Golden Age, especially in the early 1900s, engineers aspired to be "benefactors to mankind."[6] Flush with pride over massive successes—canals, bridges, dams, and public infrastructures—engineers came to believe that their profession would lead to democratically distributed prosperity for all of mankind. "Since the lot of the common man had traditionally been one of unrelenting hardship," Florman writes, "engineers during the golden age

> looked upon their works as man's "redeemer from despairing drudgery and labor." Once the common man was released from drudgery, the engineers reasoned, he would inevitably become educated, cultured and enobled, and this improvement in the race would also be to the credit of the engineering profession. Improved human beings, of course, would be happier human beings.[7]

Hence, turn-of-the-century engineers aspired to take their skills in practical application of scientific knowledge and improve the lot of humanity. Arguably, they did—for example, new sanitation systems contributed greatly to the health of urban environments.[8]

Given the seemingly boundless power of engineering and the flexibility of its central approach of applying scientific knowledge to practical problems, perhaps it is no surprise that the label "engineer" began to be used beyond the civil domain. At the beginning of the twentieth century, we see specializations of the field: electrical engineering, municipal engineering, sanitary engineering, and industrial engineering, to name a few.

All engineering is social, of course, affecting how societies function. Some subfields of engineering are more directly targeted at human action than others. Sanitary engineering, for example, affected habits of waste and consumption in urban centers. Through a seemingly politically neutral process of building infrastructures such as sewage systems, engineers started to see how their work could directly shape society. As Florman notes,

> If engineers could solve problems by being open-minded and free of prejudices—by applying scientific methods—could not all men [sic] learn to think in this mode, and then would not ignorance, superstition and bigotry vanish? "We are the priests of the new epoch," an engineering leader told his colleagues in 1895, "without superstitions."

Thus, an engineering mindset, a scientifically informed vision of how to practically and neutrally shape society, emerged around the turn of the twentieth century. And along with it came a new idea: that society itself could be engineered.

The most direct expression of this engineering-of-society mindset that appeared in the early 1900s could not be clearer: the title of "social engineer." In the late nineteenth and early twentieth centuries, a new wave of experts emerged. They sought facts, spread middle-class white American values, and worked to make everything more efficient. They placed expertise and elite knowledge above mass democratic decision-making. Their use of the title "social engineer" helped them gain legitimacy by appropriating the successes of civil and mechanical engineering. They argued that scientific thinking could be applied to society.[9]

Social engineers of this period came in three varieties. There were social reformers—Social Gospel Christian activists and early sociologists—who sought to reform society. There were management theorists, who were studying ways to manage workers in industrial capitalism. And later, there were public relations specialists, who mixed the language of social reform with the elitism of management and took to new mass media to shape society as a whole. This is the group we call mass social engineers.

Social Reformers

As multiple historians have noted, late nineteenth- and early twentieth-century America was marked by anxieties: new immigrants from Eastern Europe were thought to be diluting American values. Striking workers and socialists clashed with laissez-faire capitalists, causing social unrest with competing visions of political economy. And growth in production was leading to inefficiencies in the market because of a lack of consumers.[10] Perhaps the best expression of these anxieties came in the symbol of the crowd—the irrational masses of humanity who could not govern themselves due to their overwhelming passions, hysteria, and lack of conformity. Turn-of-the-century thinkers such as Gustave Le Bon, Walter Lippmann, and Gerald Stanley Lee warned of the dangers of these crowds.[11] As Le Bon famously argued, "the divine right of the masses is about to replace the divine right of kings."[12] The crowd was supplanting traditional leaders, bringing about fears of mob rule.

As an antidote to the unruly crowd, social theorists began to explore ways to engineer a better society. This positivist vision was fueled in part by the advent of new social sciences at the turn of the twentieth century, particularly sociology, economics, and psychology, which promised to illuminate the previously messy world of human action. The basic idea was to implement the expert knowledge gleaned from sociological surveys, economic analysis, or psychological theorizing into specific social programs meant to guide the newly ascendant masses. As the president of the American

Statistical Association put it in 1937, the application of social science to social ills would be *social engineering*, just as the application of physics to bridge-building was called *engineering*.[13]

An early example of this line of thinking is found in the work of Edwin Earp, a professor of Christian sociology at the Drew Theological Seminary. Blending together Methodist theology and social science, Earp's 1911 book *The Social Engineer* promoted "greater emphasis in education upon applied science, upon those studies in mechanics and engineering that will equip men for *doing* things as well as *knowing* things."[14] Such practical application of social scientific knowledge—mixed with the moral guidance of Christian theology—would alleviate a range of social problems, including class conflict, racial strife, "woman and child labor," divorce, "gamblers versus the people," and above all, unemployment.[15] For Earp,

> Social engineering means not merely charities and philanthropies that care for the victims of vice and poverty, but also intelligent organized effort to eliminate the causes that make these philanthropies necessary, and it means also an attempt at a readjustment of our economic and industrial system by wise statesmanship through *social control*, so that the profits of social production may be more equitably distributed to all the legitimate factors in society.[16]

"Social control" was indeed a watchword of the social reformers, a watchword that would remain central to the mass social engineers we will discuss below.

Social control through social engineering found advocates among middle-class Americans concerned about integrating the waves of predominantly Eastern European immigrants into US society. The Settlement House Movement is a key example. Found in cities such as Chicago, New York, and Boston, settlement houses were located in tenement neighborhoods populated by new immigrants who came to work in factories. In them, affluent young men and women would settle "among the urban poor, share their lot, and help them improve their lives."[17] These social reformers ran English classes, kindergartens, arts, crafts, and music classes, and discussion salons, all with the intention of "Americanizing" the immigrants.

Initially, the settlement house movement was driven by followers of Social Gospel evangelism, but over time, it became secularized and professionalized. The administrators and participants of settlement houses started college programs and began collecting data on the inhabitants of the neighborhoods they were located in, seeking out social causes for poverty and failures to assimilate into mainstream US society, and offering solutions to this problem.[18] This data collection was eventually scaled up from neighborhoods to the metropolitan level in the form of the famous *Pittsburgh Survey* or the 1919 report *Social Engineering in Cincinnati*.[19] As one historian argues, "social engineers gloried in 'facts.' Their first recommendation, no matter what the issue, was invariably the collection of information."[20] Such sociological data gathering, analysis, and intervention became a "gateway to careers in social engineering."[21] Data gathering provided a wealth of professional opportunities for social reformers and a platform for social engineering intervention in municipal politics.

During World War I, these social reformers began to target society as a whole for social engineering via government bureaucracies, arguing that their expertise would be invaluable to the war effort.[22] However, as social reformers scaled up their efforts to metropolitan or even national scale, they found that their social expertise alone wasn't enough; they needed to partner with powerful interests to implement their visions of benevolent social control.[23] They found such a partner among another set of engineer-minded people: scientific managers, who also adopted the term social engineering.

Managerialist Social Engineers

The social reformers were not the only ones using the term social engineering. So, too, were the Scientific Managers, adherents of the philosophy of Frederick Winslow Taylor, a trained engineer famous for his studies of how industrial workers did their jobs.[24] Scientific management targeted the world of industrial work, seeking to make production more efficient. As one of Taylor's disciples argued,

> As in electrical engineering we organize a field of electrical forces and resistances by arranging them into a structure of maximum usefulness, so in organization engineering we must seek to arrange a field of human forces and resistances—human motives, purposes, feelings, knowledges, and abilities—so that they interwork for maximum usefulness.[25]

Turn-of-the-century scientific managers believed that "the same laws governing the physical world governed society, so discovery of these laws would lead to the possibility of rational social control, full employment, and economic stability."[26] This is precisely the same engineering mindset found among the social reformers. Indeed, the fact that many of these managers were actually university-trained engineers only increased their credibility. And the object of social control was eerily similar: the social reformers sought to address the anxiety of assimilating new immigrants, and the managerialists addressed the anxiety brought by labor unrest—that is, unruly crowds of workers brought together in the factories.

Scientific management was, of course, the brainchild of Frederick Taylor, who sought to manage industrial workers by forcing them to operate machinery and move through space in predetermined, efficient ways—the so-called "one best way" to get work done.[27] But Taylor was not alone. His colleague Morris Cooke "expanded the domain of engineering from the study and control of materials and physical forces to the study and control of human beings. 'Social engineering' is a literal translation of [Cooke's] definition of scientific management."[28] Scientific managers like Cooke see the application of an engineering approach as "simply another indication of the passing of what may be called the 'craft spirit' in human affairs," where workers had control over the conception and execution of their jobs, in favor of "the rise of the scientific spirit," where the manager-engineers would take control over work and the worker is akin to a cog in a machine.[29] Taylor, Cooke, and their colleagues argued that such management would lead to happier and more prosperous workers, eliminating labor unrest.

Like the social reformers, the Taylorists often targeted newly immigrated Europeans for training: Taylor's most infamous analysis, for example, is his study of "Schmidt" from Holland.[30] The scientific managers also shared the social reformers' interest in gathering facts: Taylor famously used a stopwatch to measure the speed of workers' motions. His student, Frank Gilbreth (yet another engineer), filmed dozens of hours of workers in order to discover more efficient methods.[31] And, like the Social Gospel reformers, these managers saw their work as heaven-inspired. Cooke, for example, was a devout Christian. His most forceful articulation of morality and engineering was his eulogy for his mentor Taylor, which traced Taylor's ideas back to Jesus Christ: "All that Frederick Winslow Taylor, one of the greatest engineers who ever lived, did in his life time of effort was to translate into a practical, profitable, working formula, the Sermon on the Mount."[32] The managerialists also shared the social reformers' desire to spread the good word of social control throughout American society.

Managerialism even found its way into the home. Frank Gilbreth's partner Lillian (another engineer—indeed, one of the first women to get an engineering PhD) made scientific management a way of life in the home.[33] In the 1920s, Lillian Gilbreth

> engineered model kitchens—one was called the Kitchen Efficient—and purported to eliminate, for instance, five out of every six steps in the making of coffee cake. To make a lemon-meringue pie, a housewife working in an ordinary kitchen walked two hundred and twenty-four feet; in the Kitchen Efficient, Gilbreth claimed, it could be done in ninety-two.[34]

Scientific Management's entry into the home was accompanied by its entry into the communities around factories, meeting up with the social reformers who sought to Americanize the immigrant communities flocking to factory towns.[35]

Thus, the social reformer and managerialist strains of social engineering intersected in several ways. They both valued efficiency—social welfare types wanted it for governance, the managers in industry. Both were deeply affected by World War I, subsuming their specific ambitions to the war effort, seeing the war effort as a means to establish the importance of expertise in managing society.[36] For the social welfare school, the war was a chance to help, as one historian puts it, "make men moral."[37] For the managerialists, it was an opportunity to further implement their schemes during wartime industrial production and thus keep unruly crowds of laborers under control. And both sought to manage the seemingly unruly crowds of new immigrants who were coming to America around the turn of the twentieth century.

However, despite their societal-scale ambitions, both the social reformers and managerialists' scope were limited to their specific domains. The social reformers operated through bureaucracies, often butting heads against politicians and old-money aristocrats. The managerialists were more successful—after all, they were working with powerful industrialists—but their scope was limited to the workplace, and they too butted heads with government regulators who were leery of big industry.[38] The fullest expression of societal-scale, mass social engineering as a program of social control would take its final shape among a new profession that emerged in the 1920s, drew on the ideas of the social reformers, served the same industrial capitalists the managerialists served, and took as its vehicle the new communication technologies of the day. That profession was public relations, a field dedicated to the "engineering of consent." These were the mass social engineers.

Public Relations and the Mass Social Engineers

Mass social engineers owe their livelihoods to the electrical engineers who brought about new, electronic mass media in the late 1800s and early 1900s. Telegraphy, radio, cinema, and later television, along with the older technology of newspapers, all created conditions of possibility for coordinated, nation-wide media campaigns:

> With the emergence of the mass media as a connective tissue of modern life, things were happening to the texture and dissemination of information that were in the process of altering the physics of perception, changing the ways that people saw, experienced, and understood the material world and their place within it.[39]

Very quickly, the ambitions of previous social reformers and managerialists to shape society as a whole seemed possible: mass media could reach more people than just those in settlement houses or on machine shop floors. Social control on a national level appeared within reach.

The first inklings of this newfound power came from the press agents and publicists of the turn of the twentieth century. Social reform–minded journalists, dubbed "muckrakers," "utilized the power of [the] new mass media to cause a political revolt against the continued abuse of the public interest by ruthless businessmen."[40] The businesses under attack, especially railroad corporations, fought back by hiring publicists who would provide industry-friendly news stories to newspapers and magazines and shift the tide of opinion in their favor. However, these efforts were often clumsy, leading to further backlash against railroad corporations.

The clumsiness of the early publicists quickly gave way to a more disciplined—indeed, engineering-like—approach in the form of a new field of public relations, led by Ivy Lee, Doris Fleischman, and Edward Bernays, people who would "lift the lowly trade of press agentry to the euphonious heights of counselor in public relations."⁴¹ Their new field of public relations borrowed the language of the social reformers and managerialists.

> It is . . . remarkable the extent to which the first generation of PR men [sic] described their work using the Progressive idiom of their time. As many Progressives were gravitating toward techno-rational models of "expertise" and "engineering" in the quest to manage the chaos of industrial life, early public relations men also assumed a technocratic visage.⁴²

One of the borrowed terms was "social engineering." For example, in a chapter in *The Engineering of Consent*, Bernays describes public relations as "a broad social-engineering process."⁴³ Like the social reformers and managerialists, the mass social engineers recognized the rhetorical power of claiming to do "engineering."

They also shared the anxieties the social reformers and managerialists had about social upheaval and crowds. As public relations pioneer Ivy Lee argued in 1915, "this is a period of great unrest. Many strange economic and political theories are being preached."⁴⁴ Such times call for elite experts. "The crowd craves leadership," Lee argued. The experts must lead, he argued, because demagogues would do so if they did not: if the crowd "does not get intelligent leadership, it is going to take fallacious leadership."⁴⁵ If that is so, Lee reasoned, then public relations professionals, working on behalf of the nation's social, industrial, and political elites, ought to become the masters of crowds.

Lee's thinking was heavily influenced by crowd theorists like Gustave Le Bon.⁴⁶ And he wasn't the only Lee concerned about crowds. His second cousin, clergyman Gerald Stanley Lee, argued in his 1913 book *Crowds: A Moving-Picture of Democracy* that the job of the public relations professional is "news engineering."⁴⁷ For Gerald Lee, the news engineer could control unruly crowds and rise to power: "The Secretaries of What People Think, and the President of What People Think—the engineers of the news in this nation—will be the men [sic] who govern it."[48] The news engineer would be a "crowd-man" capable of leading the masses. Gerald Lee's arguments appear as a more forceful version of the Christian sociologist Edwin Earp and mix in the elitism of Taylor and the Scientific Managers.[49] His innovation was to turn to the newspaper—the mass medium capable of shaping society if only a conscious news engineer would lead the way.

Gerald Lee's theory of the news engineer was put into practice by his cousin Ivy Lee, who "demonstrated that the ambitious dream . . . to engineer 'the grandiose unification of the public mind,' was in the process of finding a practical facilitator."[50] Like the social reformers and managerialists, Ivy Lee and his fellow public relations pioneers took on the task of elite leadership, teaching crowds American values through social control. In addition, public relations adopted the social reformers and managerialists' love of facts. As Lee argued,

> We should see to it that in all matters the public learns the truth but we should take special pains to emphasize those facts which show that we are doing our job as best we can, and which will create the idea that we should be believed in. We should get so many good facts, so many illuminating facts, before the public that they will not magnify the bad. There will always be some bad facts in every business, as long as human nature is frail.[51]

This love of facts translated to an engineering approach to public relations, what Fleischman and Bernays called the "engineering of consent."[52] A husband and wife team who began a successful public relations firm together in the 1920s, Bernays and Fleischman argued that consent engineering could take place via communication technologies, particularly newspapers and radio.[53] Their use would be guided by the facts gathered from the emerging social and psychological sciences to understand and target "the group mind."[54] With "the aid of technicians . . . of communication" deploying the cutting edge, social scientific data collection and analysis techniques of the day—e.g., polls, surveys, interviews, and statistics—they believed that political leaders would be able to achieve the engineering of consent for their programs and to do so "scientifically."[55] Knowledge of the group mind would allow consent engineers to move beyond the techniques of "the old-fashioned propagandist," who was not versed in science, to control crowds through systematic engineering.[56]

Like Lee, Bernays cautioned that we must recognize that the emerging tools and techniques of mass communication could be used for good or evil, to promote or to subvert democracy, and that, as a result, "mastering the techniques of communication" for promoting socially constructive ends would be necessary for the maintenance of democratic societies.[57] If done right, the consent engineers can become an "invisible government . . . the true ruling power of our country."[58]

Lee's news engineering and Bernays and Fleischman's engineering of consent are the fullest expressions of what we call mass social engineering: the implementation of social science knowledge for the purposes of controlling the crowd. Such a mass social engineering approach echoed the Christian social reformer Earp's earlier call for "*doing* things as well as *knowing* things," defining the practice of mass social engineering as "action based only on thorough knowledge of the situation and on the application of scientific principles and tried practices to the task of getting people to support ideas and programs."[59] Mass social engineering is every bit as practical as bridge building: "Just as the civil engineer must analyze every element of the situation before he builds a bridge," Bernays wrote, "so the engineer of consent, in order to achieve a worth-while social objective, must operate from a foundation of soundly planned action."[60] And like any engineer, the mass social engineer has to apply science: as Bernays often claimed, his foundational book *Crystallizing Public Opinion*, written in 1923, was meant to be a practical manual to apply the theoretical ideas of communication theorist Walter Lippman, who wrote his book *Public Opinion* the year prior.[61]

Fleischman clarified the engineering approach, arguing for a methodology that public relations professionals continue to use to this day: research, plan, communicate, evaluate.[62] In a 1935 speech, Fleischman called for the women's fashion industry to adopt this method for a more efficient propaganda program that could communicate the latest fashions with "engineering exactness."[63] "With this as a basis," she informs her audience,

> you will set the keynote for the public, you will eliminate waste . . . and enable yourself really to avail yourself of the tools and techniques of propaganda without loss, with fullest efficiency and ultimately with the wholehearted approval of the public and the individual industries.[64]

Thus, much like the social reformers who gathered data on their target neighborhoods, or the managerialists with their efficiency-minded work motion studies, Lee, Bernays, and Fleischman prescribed a method for mass social engineers: get the facts, study the public, discern psychological ways to influence them, and communicate with them, ideally by creating newsworthy events.[65] (Indeed, as we will show in the next chapter, this basic pattern will reappear in interpersonal hacker social engineering.)

But more so than social reformers or managerialists, their approach was expansive. Mass social engineering had wide applications across every domain of American life, from consumption (e.g., Fleischman's recommendations to the fashion industry) to support for industry (e.g., Lee's work for the railroads) to support for war efforts (e.g., Bernays's work as part of the World War I Creel Committee). This was crowd mastery on a large scale. Overall, if the mass social engineer is successful, Bernays famously argued "the ideas conveyed by the words will become part and parcel of the people themselves."[66]

Like the social reformers before them, Bernays, Fleischman, and their colleagues presented themselves as politically neutral, "detached manipulators" of the masses.[67] Mass social engineering had to appear to be apolitical to be effective:

> For self conscious "professionals" such as Bernays, expert detachment was, of course, a point of pride—and a strong selling-point, as that hard attitude provided tacit reassurance to potential clients that the propagandist worked not just by instinct or mood, but as impartially, and, if need be, ruthlessly as any doctor or attorney.[68]

Or, we might add, as impartially and even ruthlessly as an engineer.

"Social Engineering" Becomes a Pejorative

However, mass social engineering has long since become a pejorative. Indeed, as the engineer Samuel Florman noted in 1976, by the middle of the twentieth century into the 1970s, *all forms* of engineering, from civil to social, came under attack.[69] The Golden Age of Engineering had come to an end. For its part, by the tail end of the twentieth century, mass social engineering was seen as a failure at best and the path to totalitarianism at worst, dismissed as mere propaganda or feared as a cynical attempt to control the public. Florman concluded his chapter on the fall of engineering with engineering's most glaring failure: its failure to apply "right reason" to social problems:

> As for the ultimate hope, that the engineer's rational thinking would show the way toward solutions of society's problems, the unanticipated events of each incredible, tumultuous day demonstrate convincingly what a naïve conceit that was. . . . In retrospect, the ideals and dreams of engineering's Golden Age seem foolish and immature.[70]

Part of the reason mass social engineering came under attack was that, despite the claims to neutrality and the professed love of facts, mass social engineering was quickly exposed as a deceptive, manipulative practice in service of self-interested elites. Even the most sanguine history of public relations, such as public relations scholar Scott Cutlip's *The Unseen Power*, documents many occasions where mass social engineers hide their advocacy for the powerful clients behind front groups and omit facts—or even outright lie—in order to present their clients in the best possible light.[71] Combine these deceptions with the elitist conceit that a privileged few needed to master the "crowds" of American democracy, and there are plenty of reasons to decry mass social engineering.

Indeed, by the 1970s, the term "social engineering" took on the pejorative meaning it keeps with it to this day, particularly for thinkers on the American political right reacting to perceived excesses of Progressivism as well as Lyndon B. Johnson's Great Society. As early as 1955, William F. Buckley Jr.'s new *National Review* magazine began with "Our Mission Statement," which noted that, "The profound crisis of our era is, in essence, the conflict between the Social Engineers, who seek to adjust mankind to conform with scientific utopias, and the disciples of Truth, who defend the organic moral order."[72] This vision of the conservative, moral Truth versus the liberal Social Engineers is mainly aimed at the social reform element of social engineering, not necessarily the mass media practices of people like Bernays.

But conservatives also found fault with what they saw as mass mediated social engineering. An example is Nixon campaign strategist and conservative columnist Kevin Phillips's book *Mediacracy: American Parties and Politics in the Communications Age* in 1975.[73] Throughout *Mediacracy*, Phillips expanded Buckley's thesis. Phillips first describes the "knowledge elites"—"media, educators, and city planners"[74]—who seek to use their social scientific knowledge to engineer society via the new communication technologies to match their Great Society vision, especially in terms of sexual revolution or income redistribution to what Phillips called the "ghettos." Against this knowledge elite, Phillips counterposed the vast "middle class" who are deeply frustrated from such top-down control and its attendant taxation, who (he claimed) merely want free exchange, open markets, and no discussion of race or racial reparations and definitely no deviant sexuality. Such opposition to racial justice or sexual liberation for women or queer people continues in American conservative thinking to this day. For example, in a 2010 *Wall Street Journal* profile, conservative publisher Emmett Tyrrell contrasts the conservative vision with the liberal:

> Mr. Tyrrell finds liberals' attitudes to be as vexing as their policies: "There is only one political value that they have stood by through three generations, and that is the political value of disturbing your neighbor." If conservatism is a temperament, he adds, "liberalism is an anxiety—an anxiety about life, liberty and the pursuit of happiness, which explains their eagerness to coerce, to tax, to social-engineer."[75]

However, despite conservative articulations of "social engineering" with left politics, it's not fair to suggest that the political left uncritically supported mass social engineering. For example, in 1959, the sociologist C. Wright Mills published *The Sociological Imagination*, which included a scathing indictment of quantitative social sciences:

> Among the slogans used by a variety of schools of social science, none is so frequent as, "The purpose of social science is the prediction and control of human behavior." Nowadays, in some circles we also hear much about "human engineering"—an undefined phrase often mistaken for a clear and obvious goal. It is believed to be clear and obvious because it rests upon an unquestioned analogy between "the mastery of nature" and "the mastery of society."[76] [The human engineers'] political philosophy is contained in the simple view that if only The Methods of Science, by which man now has come to control the atom, were employed to "control social behavior," the problems of mankind would soon be solved, and peace and plenty assured for all.[77]

For Mills, such social engineering (or "human engineering," as he calls it) reflects a deadening bureaucratic mindset in which technocratic elites reduce all human behavior to whatever can be measured.

By the 1970s, thinkers on the left like Stuart Ewen decried the efforts of Bernays, Ivy Lee, and other "captains of consciousness," blaming them for more than just the deadening reduction of humans to atoms.[78] Ewen's major concern is with the destructive rise of consumerism.[79] As Ewen argues, "consent engineering" was part and parcel of a larger transformation of American society from a largely self-sufficient agrarian society into a mass culture of consumption, where every facet of life was conditioned to support capitalism. After industrialists had disciplined workers to meet the needs of factories in the nineteenth century, the glut of products that emerged in the twentieth century compelled capitalists to "manufacture customers" to consume those goods.[80] Workers were compelled to produce in order to consume, and the social reforms of the progressive era gave way to confusing the ownership of cars and washing machines with the possession of a good life. But, by the time of his analysis in 1976, "Americans have increasingly questioned" the consumerist logics of consent engineering.[81] Ewen argued that the social movements like the civil rights, students', and women's movements had to slough off the deadening weight of mass culture, the very culture the engineers of consent were seeking to foist on the nation.[82]

Even the field of public relations itself began to reject social engineering as an approach. In 1962, a public relations practitioner was not too subtle in his critique of Bernays and Fleischman's "engineering of consent":

> The "engineering of consent" implies the use of all the mechanics of persuasion and communication to bend others, either with their will or against their will, to some prearranged conclusion, whether or not their reaching that conclusion is in the public interest. I can't

help but think that carrying the "engineering of consent" line to a logical conclusion is the pistol at the back of the neck, reminiscent of Nazi times and not unknown behind the Iron Curtain.[83]

As Florman lamented in 1976, the connotation of "engineer" had been torn from its associations with the modernist realm of progress and had been rearticulated with visions of jackbooted government agents and the deadening conformity of marketers, both controlling crowds through mass mediated messages. Even public relations professionals—colleagues of Lee, Bernays, and Fleischman—decried the very idea of engineering society.

Conclusion

Overall, then, elite technocrats' implementation of social science knowledge to engineer society at the mass scale via communication technology went from a modernist vision to a fearsome nightmare. By the time Ewen interviewed Bernays in the 1990s, Bernays's elitist attitudes were passé, a holdover from a time when engineers ran the world.

And yet, we should note that the underlying logics developed in the 1920s by mass social engineers such as Bernays, Fleischman, and Lee are still very much with us. Public relations, after all, is still a major occupation. These days, it sometimes goes by different monikers: *strategic communication* or *political communication*—perhaps to shake off the old connotations of "public relations," just as "public relations" was meant to supplant the older term "propaganda." Despite these name changes, the contemporary fields of strategic communication and political communication often echo the older mass social engineering ways of thinking. Consider this contemporary definition of political communicators as people who

> build political consensus or consent on important issues involving the exercise of political power and the allocation of resources in

society. This includes efforts to influence voting in elections as well as public policy decisions by lawmakers or administrators. On the international level, this includes communications in support of public diplomacy and military stabilization.[84]

"Building consent." Perhaps consent is no longer to be *engineered* by powerful social actors—whether corporate or governmental—but is instead meant to be *built*.

Moreover, social engineering with mass scale ambitions has returned in a strange new form, the one we're calling "masspersonal social engineering," mixing mass with personalized messaging. Via the new corporate social media, powerful actors can turn "political communication into an increasingly personalized, private transaction . . . , [which] fundamentally reshapes the public sphere, first and foremost by making it less and less *public* as these approaches can be used to both profile and interact *individually* with voters outside the public sphere."[85]

But to understand how mass social engineering can become a masspersonal form—targeting individuals but with an eye towards shaping entire swathes of society—we have to consider the other meaning of "social engineering" that emerged just as mass social engineering was being critiqued in the 1970s. That form is *interpersonal social engineering*, a form developed by phone phreaks and hackers. We turn to them next.

2
Phreaks and Hackers: The Rise of Interpersonal Social Engineering, 1976–Present

> Engineering is when you take a wrench to a bolt, turn it, and something happens. Social engineering is when you take a telephone, get somebody who's dumb as a bolt on the other end, and turn it.
>
> —Chesire Catalyst[1]

Although mass social engineering was widely vilified by the 1970s, the idea of social engineering itself did not die out at that time. While Americans were decrying the idea that elite technocrats wanted to engineer perceptions on a mass scale, a new, more personal form of social engineering emerged. It will be most associated with computer hacker and phone phreak cultures.

In many ways, this new form appears to be quite different from the older, mass form. Whereas mass social engineers sought to master "the crowd" via mass media messaging, phone phreaks and hackers had a very different goal in mind: access to telephone and computer networks, respectively. While the mass social engineers wanted to shape society as a whole, the phreaks and hackers engaged in quick, instrumental manipulations of telephone operators or computer lab administrators—their targets were individuals and their approach was more akin to interpersonal communication than mass communication. And while the mass social engineers drew from the new social sciences of their day, the phreak and hacker practice was—at least at first—a jocular, instinctual, underground practice.

But the mass and interpersonal hacker forms of social engineering have many connections. As cybersecurity scholar Joseph Hatfield argues, there are several conceptual and semantic overlaps between mass and interpersonal social engineering.[2] Moreover, masspersonal social engineering, a synthesis of these two forms of social engineering, is with us today.

Intellectual Roots of Hacker Social Engineering

A new form of social engineering emerges from the fears of the older mass social engineering, particularly the fears of mass manipulation carried out by government or corporate actors. Mass culture was—depending on one's political predilections—either a symptom of governmental overreach in people's lives, or a symptom of the deadening conformity and bankrupt values of consumer capitalism.

By the mid-twentieth century, social movements and intellectuals sought to move beyond mass culture by championing new forms of individualized technologies and techniques. This was the period the American novelist Tom Wolfe famously dubbed the "Me Decade" in 1976.[3] The turn to the individual appeared in three forms. First was New Communalist thinking, starting with growing interest in personal technologies in the 1960s carrying through to contemporary internet culture. Second was the turn to self-help and individual therapeutic cultures. Third was the hyper-individualism of neoliberal economics. All three would go on to inform interpersonal social engineering.

First, communication scholar Fred Turner's excellent profile of Stewart Brand in *From Counterculture to Cyberculture* illustrates the turn to personal, liberatory technologies.[4] Brand was a quintessential New Communalist, a leader of the back-to-the-land movement of the 1960s. While "back to the land" implies a renewed pastoralism, Brand's vision brought together "the world of university-, government-, and industry-based science and technology; the New York and San Francisco art scenes; the Bay area psychedelic community; and the communes that sprang up across America in the 1960s."[5] Here, the doors of individual perception could be opened up through LSD or through personal electronics. The heterogeneity of such a mix is illustrated by any given page of Brand's famous *Whole Earth Catalog* counterculture bible. Any given page of the catalog might have a Native American-inspired buckskin shirt, a book on psychology, and a microcomputer. As Turner argues, Brand's communalism was not just about the land and a new pastoral vision; it embraced "the notion that small-scale technologies could transform the individual conscious and, with it, the nature of community."[6] Brand's New Communalist theory of individual technologies and individual consciousness-raising would later find a home in 1990s cyberlibertarian thinking about the internet, culminating in the birth of *Wired* magazine and online discussion forums such as the Whole Earth 'Lectronic Link or WELL.

Alongside the New Communalism of Brand in the mid- to late twentieth century, we also see a rise in individual-scale techniques of self-help therapy culture. This culture included psychoanalytic and psychiatric techniques aimed to help individuals overcome subconscious conflicts. But the therapeutic culture of the 1970s was not purely about trained psychological experts aiding individuals; it also found expression in wildly popular self-help texts. Thus, the technical expertise of therapy was mediated by the rugged individualism of American culture and consumerism: "the language of [Freudian] psychotherapy left the realm of experts and moved to the realm of popular culture, where it interlocked and combined with various other key categories of American culture, such as the pursuit of happiness, self-reliance, and the belief in the perfectibility of the self."[7]

One such individualized field of therapy that arose in the 1970s was called neuro-linguistic programming (NLP). Developed by the therapists Richard Bandler and John Grinder, NLP is a therapeutic technique that involves hypnosis and suggestion in order to help patients alleviate phobias and overcome psychological problems.[8] Like its name implies, NLP practitioners argue that a patient can be reprogrammed by having their memories of past experiences modified through suggestions and connections to other, more positive memories. "NLP was cast in the mold of ascendant technology: language was a way of programming the neural machine."[9] Above all, NLP holds that the individual—not broader social structures—is fully responsible for their own phobias, fears, and problems. Reprogramming the self takes precedence over tackling social ills.[10] Bandler and Grinder's 1979 book *Frogs into Princes* allowed individuals to study and apply this ascendant technology to their own lives.[11] Self-improvement is a far cry from the societal-scale manipulation envisioned by mass social engineers. (Incidentally, NLP was quite influential among hacker social engineers; we will return to NLP in chapter 6, "Penetrating.")

The individualized techniques of self-help and therapy influenced a new form of individualized, self-help politics.[12] A key text is Mark Satin's book *New Age Politics*, written in 1976 and published commercially in 1979. *New Age Politics* echoes the fears of mass social engineering, condemning "'monolithic institutions,' or institutions with totalizing, controlling power: transportation, medicine, schooling, religion, the nuclear family, nuclear power, the defense system, the monolithic state, the governing elite, and so on."[13] Satin counters the monolithic with "personal responsibility, self-reliance, freedom of choice, belief in ethical values and ideals, and an all-encompassing love for what some of us choose to call God."[14] Here, a politics driven by technocratic elites is replaced by the therapeutic politics of self-consciousness.

As for economics, we can also point to the well-documented rise of neoliberalism in the 1970s.[15] Neoliberalism was a return to the economic liberalism of Adam Smith's 1776 treatise *The Wealth of Nations*, which proposed that society is directed by an "invisible hand" that emerges from the aggregated, market actions of rational individuals.[16] This return to individualism in economics repudiated strong governmental regulations of the economy—not to mention anything that hinted of socialism—in favor of a reduction of all human action to individual, rational choices in a free marketplace. Neoliberalism found homes both in the mainstream schools of economics, notably at the University of Chicago's Milton Friedman and to a lesser extent schools in Virginia with Gordon Tullock and James Buchanan, as well as countercultural movements such as Agorism.[17]

This historical context—the rise of individualism in the 1970s—has two aspects relevant to our interest in social engineering. First, the rise of the individual hammered home the critiques of mass social engineering prevalent from the 1920s through 1950s. Society-wide solutions, theorized by social scientists and implemented by elite, technocratic mass social engineers, were out. Individual-scale technologies and techniques, self-help therapeutic thinking and politics, and the liberation of one-to-one economic exchange, were in.

Second, within this milieu, social engineering arose again, albeit in a very different form from the older mass form. Sometime in the mid-1970s, a group of people who called themselves "phone phreaks" started talking on the telephone about their own version of social engineering.[18]

The Phone Phreaks

"In the mid [1960s]," writes a journalist in 1971, the Bell Telephone company,

> and almost every other phone company in the world, decided to convert billions of dollars of telephone equipment to a compatible system based on 12 tones. Ever since, groups of college students high on electronic circuitry, freckle-faced teen-agers and blind kids with sensitive hearing have kept up with the latest telephone technology. They call themselves "phone freaks."[19]

The phone phreaks (two *ph*'s are preferred) were deeply enamored with the Bell Telephone system. The phone network encompassed, as the old AT&T advertisements put it, "one policy, one system, and universal service"—a massive, continent-spanning, monopolized technological system, and the phreaks wanted to explore it all.[20]

Phone phreaks were found all over the United States and include a cast of colorful characters, many of whom will appear in the subsequent pages of this book: John "Cap'n Crunch" Draper, an Air Force-trained electrical engineer; Denny Teresi, a blind teenager from California; Susan "Thunder" Headley, a rock groupie, sex worker, and coin collector; Kevin Mitnick, who would infamously lead the FBI on a manhunt in the 1990s; John "Corrupt" Lee, a Black kid in New York City who split his time between a street gang and a prestigious private high school; and Jordan Harbinger, a cell phone phreak who would later become a pickup artist.

As historian and technology entrepreneur Phil Lapsley argues in his book *Exploding the Phone*, the phreaks' obsession with exploring the phone network required that they either pay hefty long-distance phone bills or find ways to avoid paying.[21] They often chose the latter, and thus the phreaks were vilified as fraudsters by the phone company and various government agencies. But as the phreaks often emphasized, they weren't simply scamming the system. They had an intense desire to connect with one another, to gain knowledge of the furthest reaches of the Bell system, and, of course, to be recognized by their peers for this knowledge. While Steve Wozniak is best known as a co-founder of Apple Computers, he was also a phone phreak. As he notes in his forward to *Exploding the Phone*,

> for me, phone phreaking was a place in the world that I was like a leader. It was a place where I could blossom. And it's not that I could blossom as a criminal—it wasn't that we had lots of people to call or had giant phone bills or really wanted to rip off the phone company or anything. It's just that it was so exciting! When I went into a room and showed off phone tricks with a blue box, I was like a magician playing tricks. I was the center of attention. That was probably partly what drove me. But it was also the fascination of doing something that nobody would really believe was possible.

And as Susy "Thunder" Headley put it in her testimony to the US Senate in 1983, when phreaks got together, "everybody was trying to outdo everybody else to see who was greater, who was better, who had better access, who could get into the tightest or best systems."[22]

Thus, for many phreaks, getting free long-distance calls was a means to a larger end: to see where the phone can take you, to understand how its connections are made, to find out if you can connect multiple people on a single call, or simply to hear the various sounds of switching equipment around the world—and to show off your feats to others.[23] Lapsley's *Exploding the Phone* describes the Bell Telephone system as a "giant cyber-mechanical-human system . . . , the largest machine in the world."[24] For the phreaks, the Bell System was a technological playland. Their ability to explore it gave them social capital and influence among their social circles.

Phone phreaks were proto-hackers, playing with phone technologies to see what could be done with them. Indeed, most histories of hacking find that phreaking led directly to computer hacking.[25] This is not surprising when we look at phone phreak publications. Early magazines, such as *Telephone Electronics Line* and *TAP* in the 1970s and 1980s, were largely dedicated to the phone system. A publication that followed on their heels, *2600*, built on their foundation—in fact, the name "2600" refers to the frequency in Hertz of a telephone control tone. But *2600*, which launched in 1984 and is still in print to this day, transitioned from focusing on phone phreaking to computer hacking. This transition in focus can also be seen in some of the key figures in this history—Susan Headley, Lewis De Payne, and Kevin Mitnick all self-identified as phone phreaks in the 1970s and '80s but eventually reidentified as hackers by the 1990s. Moreover, many phone phreaks shared their knowledge of their craft on 1980s-era computer bulletin board systems (BBSs), a practice that required knowledge of phones, computers, and networking.

Phone phreaks and their hacker offspring certainly cultivated technical knowledge and technological mastery of their respective target systems. They developed personal liberation technologies, such as Blue Boxes that could create telephone control tones, or Red Boxes that could imitate the sounds of coins clinking into payphone slots. Their hacker offspring celebrated the personal computer as a liberation technology. They also aped some of the language of the counterculture, thinking of phones and computers as windows into new cyberworlds. They often described their activities in self-help terms, claiming to seek knowledge to improve their skills. And they shared some of the distaste of the neoliberals with the phone company, which was a state-sanctioned monopoly up until the Reagan administration. As Headley put it, "the phone company is a monster that should have been killed years ago."[26]

They had a social side, as well. This might appear in hours-long (or even days-long) "conference calls" or in intense online socializing over Internet Relay Chat or BBSs. Or it would appear in conventions where they get together to talk about telephones or computers. But most importantly for us, phreak and hacker sociality also appeared in what they called "social engineering," a term they started using in the mid-1970s.

Phreaking and Hacking via Social Engineering

Phone phreak social engineering is, at its core, very simple. Rather than using a piece of technology such as a Blue Box to control the network, often it was easier for a phreak to call up a telephone operator and simply ask for access to restricted parts of the network. As one phreak explained in 1973, "some of the best phone phreaks don't use any hardware at all. You can talk people in the telephone company into doing stuff for you."[27] A phreak social engineer might use the pretext that they are a customer in need of a service, or more often the pretext of being a fellow employee in need of access to special technical test lines or capacities. The documentary *The Secret History of Hacking* recreates one of these engagements. In this recreation, phone phreak Dennis Terry offers a typical conversation: "How ya doin'? Good, buddy! This is Bob from the Alpine Office in Phoenix, Arizona. We have a test that we're needing to run, a transmission test . . ."[28] This friendly banter—coupled with phone company lingo—tricks the telco employee into believing they are talking to a fellow employee. Access is soon granted. After they get access, the phreak can then make their long-distance calls or set up conference calls with fellow phreaks. These practices carry forward into the world of computer security hacking: often, it's easier to just call or email someone and just ask for a password than it is to break through encryption.

But why call something that sounds like con artistry "engineering"? Why use such a respectable word for such trickery? The mass social engineers took up the term to invoke a science-to-implementation model and emulate the triumphs of nineteenth- and twentieth-century civil engineering. In contrast, phreaks were making a joke. Indeed, a more common term they used for conning a telephone operator seems far more honest: "bullshitting." A hacker called The Mentor sums up the relationship between the terms: "For those of you who may be new, 'engineering' is short for 'social engineering,' which is long for 'bullshitting.'"[29]

But eventually, "social engineering" takes over as the name of this hacker practice. There's nothing in the hacker literature that explains why the term "social engineering" came to predominate. However, given that the phreaks were interested in the vast telephone system—a feat of electrical engineering—and given that they included in their ranks trained electrical engineers like John "Cap'n Crunch" Draper, the playful idea that manipulating people was an engineering process makes sense. This term carried on as phreaking transitioned to computer hacking in the 1980s.

Moreover, the phreaks accidentally hit on a curious etymological feature of the word "engineering": its root is *gin*. From that root we get the term "ingenuity" and its related words "engine" and "engineering," which refers to the capacity to manipulate one's environment. We also get an obsolete—yet quite illuminating—meaning: *gin* is defined in the *Oxford English Dictionary* as "a snare, net, trap."[30] This included *verbal* traps. Indeed, in many uses in English history, "engine" actually referred less to physical technologies and more to the manipulative machinations and tricks of villains, culminating in a villainous character called "Malengin" in Edmund Spenser's epic sixteenth-century poem *The Faerie Queene*.[31] Malengin, it turns out, is quite adept at manipulating people with his use of language.[32] Had Spenser been writing in the 1970s, Malengin would probably have thrived among the phone phreaks.

Although technologies have changed over the decades since the 1970s, this interpersonal hacker form of social engineering is still with us today. Landline phones gave way to wireless telephony; simple passwords gave way to two-factor authentication and strong encryption. Despite these changes, simply tricking someone into giving up access to a network or resource remains effective. We might think of hackers as hoodie-clad loners hunched over glowing keyboards, furiously typing away to break through encryption. We may think of them as anti-social, just as incapable of understanding humans as they are capable of understanding computer networks. The hacker practice of social engineering, however, reveals that they are quite capable of relating to others. And while we may think of technical attacks as the key vector by which computer networks are compromised, it turns out that social engineering—*gins*, interpersonal nets and traps—is very common and extremely effective. A key annual report in the cybersecurity world, the *Verizon Data Breach Investigations Report*, consistently lists social attacks as the most common and dangerous vector by which hackers gain access to restricted networks.[33] As the social engineer Kevin Mitnick is fond of saying, in information security, "the human is the weakest link."[34]

Hacker Social Engineering as Interpersonal Communication

Phreaks, and the later hackers, did their work through interpersonal communication channels. The phreaks' preferred medium, of course, was the telephone. As historian Claude Fischer explains, early telephony was an invention seeking a use. Initially, telephone companies tried using it as another broadcast medium, transmitting "news, concerts, church services, [and] weather reports."[35] Eventually, however, telephone users themselves transformed it into a medium for sociality by the 1930s.[36] Pairs of people, or small groups on party lines, built and maintained intimate relationships over the phone. By the 1950s, the telephone was understood to be a medium for such intimacies. As Bell engineer A. B. Clark writes in 1952, "The value of a connection by telephone cannot be measured by the number of words spoken for it brings people together in great intimacy so that understandings can be reached almost as well as in a direct meeting."[37]

The phone phreaks were like anyone else; they built relationships over the phone. Since phreaks were all over the country—indeed, all over the world—they had to find a way to talk to one another without paying expensive long-distance bills. They used social engineering, in part, to enable their interpersonal communication. And their social interactions, in part, helped them practice the interpersonal communication skills that enabled them to more effectively social engineer telephone company employees.

The interpersonal aspects of phone phreak social engineering have translated into other areas. Contemporary, hacker social engineers use email or texting, two other media that interpersonal communication scholars recognize as useful for building interpersonal relationships.[38] Social engineers even do their work face to face—the gold standard of interpersonal communication—when they do "physical penetration" of organizations. They go in person to targeted organizations and persuade employees to give them access to restricted areas.

Thus, one major contrast between the early twentieth-century form and the 1970s phone phreak form of social engineering hinges on the target: the crowd versus the individual.

Relating Mass and Interpersonal Social Engineering

So far, we have illuminated two distinct meanings of social engineering: the mass form, which was developed in the early twentieth century and found its greatest expression in the "consent engineering" of Lee, Bernays, and Fleischman, and a playful, interpersonal form used by phone phreaks to con telephone operators into giving them access to restricted parts of the telephone system.

If what the phreaks and hackers are doing is "social engineering," it appears to have little relation to the older, mass form. However, as US Naval Academy professor Joseph Hatfield argues, the two practices share many semantic and practical overlaps: both rely upon what he calls *epistemic asymmetry, technocratic dominance,* and *teleological replacement*. Put simply, these terms mean that all social engineers hide their true purposes, use sociotechnical knowledge to control others, and seek to manipulate others into doing things they may not necessarily normally do.[39]

We can see further overlaps between the two forms of social engineering when we look at contemporary definitions of the terms. First is a definition of what we would call mass social engineering, coming from political theorists:

> Social engineering means arranging and channeling environmental and social forces to create a high probability that effective social action would occur. The word engineering suggests the designing and erecting of structures and processes in which human beings serve as raw material.[40]

And here is a definition of the interpersonal hacker form of social engineering from security consultant Chris Hadnagy: "Social engineering is *any* act that influences a person to take an action that may or may not be in his or her best interests."[41] For Hadnagy, this typically means convincing people to give out their passwords, give access to restricted information systems, or reveal private or proprietary information.

Both definitions, we suggest, share a concern with the place of humans within complex sociotechnical systems. For the mass social engineers of the early to mid-twentieth century, human society could be known through social science, and that knowledge could be used to inform mass communication messages that would direct and shape society. While traditional fields of engineering had brought about impressive new sanitation, transportation, and communication infrastructures, the world of humans was still messy and in need of social control to maximize the benefits of modernity and take full advantage of the fruits of technologists. For the interpersonal hacker social engineers of the 1970s onward, individual humans who interact with technologies could be convinced to trust the hacker and provide the hacker access to restricted information. A hacker might be merely interested in exploring a system. Or, the hacker might have criminal intentions. Either way, the human operator of the system was a barrier to entry to it, and thus had to be persuaded to give the hacker access. To riff on Mitnick, for both types of social engineering, humans are a "weak link," a raw material in need of manipulation in order to achieve what the social engineer wants.

Indeed, when it comes to controlling the "weak link," both mass social engineering and interpersonal hacker social engineering share a love of expertise and elitism and a disdain for their targets. When Ewen interviewed Bernays, Bernays was not shy about proclaiming a need for an "intelligent few" to control the hapless masses through their superior knowledge.[42] Hacker social engineers can be even more blunt in their assessment of their targets. In a panel discussion at the 1994 HOPE conference, the publisher of the *2600* hacker magazine Emmanuel Goldstein quipped that social engineering "is amazingly like computer hacking, except the people at the other end aren't quite as intelligent as computers." The audience laughed heartily at that line.[43] Both variations of social engineering value technical knowledge and often see their targets as hapless rubes, demonstrating what Hatfield calls "epistemic asymmetry" and "technocratic dominance."[44]

This shared love of the technical and of technique results in a systematized approach to engineering humans within sociotechnical systems. Inspired by civil engineering, the mass social engineers took this rational approach largely as a given. As we will explore in chapter 6, "Penetrating," the hacker form of interpersonal social engineering would also eventually draw on social science, just as the previous mass form did. While the phreaks and hackers initially started out using intuition and talent to manipulate their targets, by the late 2000s, professional social engineers would develop a theoretical literature, largely culled from evolutionary psychology, to systematize their practices. In this sense, interpersonal social engineering echoes the earlier mass social engineering conception of science-to-implementation, furthering their technocratic dominance.

This brings us to a major concern of this book: a focus on, and theorization of, the broad, systematized process of social engineering writ large. Recall the processes of "consent engineering" Doris Flesichman suggested, discussed in the previous chapter: get the facts, study the public, discern psychological ways to influence them, and communicate with them. This basic pattern of research, scenario construction, and engagement also appears in the contemporary, professionalized social engineering literature. For example, the core chapters of security consultant Sharon Conheady's book *Social Engineering in IT Security*, are "Research and reconnaissance," "Creating the scenario," "Executing the social engineering test," and "Writing the social engineering report." Published in 2014, Conheady's book systematized what the phreaks and hackers had been doing since the 1970s.[45]

Indeed, as our book will explore in more detail, this mid-1970s practice of social engineering replicated—often unconsciously—many of the techniques developed by the mass social engineers of decades prior. We see in the phone phreaks' and hackers' techniques a set of concepts and practices that bridge the gap between the mass social engineering of the early twentieth century and of masspersonal social engineering. We also see convergences among all types of social engineering at the level of social science theories, especially as those theories pertain to communication.

Conclusion

Despite the fears of mass social engineering and the turn to individualism in the 1970s, the gap between the mass and interpersonal forms of social engineering is not as wide as it may initially appear. Taken together, both forms reveal a spectrum of techniques capable of targeting individuals (as the phreaks and hackers did) but with societal implications (as the mass social engineers sought to have).

As we discussed in the introduction, we see masspersonal social engineering as part of the broader phenomenon that O'Sullivan and Carr have called masspersonal communication.[46] But we have also been inspired by the methodological approach that they take in their work. They note that recent developments in computer-mediated communication, in particular the internet and social media, highlight the blurring boundary between mass and interpersonal models of communication that were at the core of dominant communication theory for most of the twentieth century. But instead of seeing that blurring boundary as something entirely new and technologically determined, they see it as an opportunity "to see if similar phenomena have existed with older technologies." Looking for historical examples or precursors to a seemingly new phenomenon provides the opportunity, therefore, to revise and "explain communicative phenomenon more precisely and comprehensively," "to reassess some basic assumptions," and, in doing so, "to advance the communication discipline by proposing revised assumptions." In their case, an examination of emerging digital communication technologies and their use allows them to "challenge one of the enduring frameworks" of communication, which is "the definitional divide between interpersonal and mass communication."[47]

Similarly, our approach in the next section will be to take up the more recent form of hacker social engineering and consider the ways in which its practices can be found in the older, mass form. Specifically, we will draw on the literature of hacker social engineers—underground magazines and online forums, podcasts, articles, and books—to illustrate steps of information gathering (trashing), scenario development (pretexting), engagement (bullshitting), and success (penetrating). But rather than treat these practices as a radical break with the past, we will use them as lenses to look backward at the mass social engineering we discussed in the previous chapter. And we will also look forward to the contemporary practice of masspersonal social engineering as marked by trashing, pretexting, bullshitting, and penetrating. These temporal motions across the past century will further bridge the gaps between all of the forms of social engineering.

II
The Social Engineering Process

3
Trashing: From Dumpster Diving to Data Dumps

> Here's the main lesson of garbology: People forget, they cover, they kid themselves, they lie. But their trash always tells the truth.
>
> —William Rathke[1]

> Stare into dumpsters long enough and they stare into YOU.
>
> —John Hoffman, *The Art & Science of Dumpster Diving*[2]

It all starts with trash.

The book *Masters of Deception*, by Michelle Slatalla and Joshua Quittner, is a story of the late 1980s New York City–based hacker group of the same name, opens with a teenager named Paul in a dumpster in Astoria, Queens:

> This is one way to become a computer hacker, the way Paul has chosen. He tries to snare one of the five or six invitingly swollen bags that sit in the bottom of the dumpster. . . . There's nowhere he'd rather be than here, rummaging around in this dark alley in a dumpster full of phone company trash, looking for computer printouts.[3]

Paul and his Masters of Deception compatriots were continuing the long and honored phone phreak and hacker practice of *trashing*—"climbing around in garbage, where you hope to find computer printouts that list secret passwords and logons."[4] Trashing has been used by phone phreaks and hackers since at least the mid-1970s, and its intricacies have been discussed in zines such as *Telephone Electronics Line, Technological Assistance Program*, and *2600*, and in computer bulletin board posts.[5]

Trashing even enjoyed a moment on national television in 1982 when Geraldo Rivera warned the United States about the "electronic delinquents" (computer hackers and phone phreaks, including Susy "Thunder" Headley) breaking into computers. As Rivera intoned to the camera during an episode of ABC's *20/20*,

> You might be asking yourself how a group of 16- or 17-year-old kids with no formal computer training got ahold of the sophisticated codes needed to give them access to the telephone company's computer system. Well, according to the kids themselves, the answer comes from right here: the trash bins located behind the phone company building. That's right, the trash bins. It was here that they found some out-of-date manuals that contained the information they needed.[6]

Rivera's claims were supported by none other than Headley herself, who testified to the United States Senate in 1983 that she and her colleagues would get passwords through "various research methods. Garbology is a polite term for it; going through the garbage of a company and finding computer printouts that have passwords on them, finding notes that people have jotted down."[7]

Trashing is a practice that hackers share with private detectives, paparazzi, scavengers, and urban freegans, all of whom see garbage cans as a valuable source of materials ranging from gently used goods to food to art supplies to information.[8] As for Paul and the hacker group Masters of Deception, they were able to use their trash-begotten knowledge to do some interpersonal social engineering on a telephone operator. The information they gleaned from that night's trashing run included a "secret list of internal New York Telephone Company numbers." With this, they could call up the New York telco business office and pretend to be

> Lou in Provisioning [and] let it drop that you know all about a certain computer at a certain phone number. You're a part of the great Bell family, and of course the business office is going to give you the information [you're asking for]. You sound like you know the person: "Don't you remember me, Barry? We've talked before, fella."[9]

Like a freegan dumpster diver finding edible food, a bicycle, or perfectly good shoes, phone phreaks or hackers are able to convert something that was devalued into something very valuable: deeper insight into and even unfettered access to a vast communications network of computers and telephones.

We start with trashing because the process of converting discarded or overlooked information into valuable knowledge is a key practice in all forms of social engineering. In order to do their job—manipulate people for purposes of social control—social engineers need information on the crowds, organizations, or people they are targeting. These days, of course, they can go online and gather Big Data. In the days of the mass social engineers, the "engineers of consent" like Bernays, Lee, and Fleischman did surveys or in-home consumer observation, drawing on techniques of the new social sciences. For example, as a 1942 article on "advertising engineering" suggests, families should be studied on a weekly basis for years in order to gain enough information about them.[10]

But data from surveys or big datasets shares many links with the down-and-dirty trashing practices of the phone phreaks and hackers of the 1970s through 1990s. The epistemological practice of turning overlooked information into social control is brought into sharp relief if we consider trash, rather than the more sterile "data," as a source of knowledge. Whether they seek to control individuals or masses, social engineers have to go through the trash. Here, we explore hacker social engineering trashing, first explaining how it came to be possible, how it relates to the older mass social engineering, and then how the trashing way of thinking has been transformed in a time of Big Data and Open-Source Intelligence (OSINT). Hackers and phreaks show us that surveys, Big Data, online research, and garbage are all more alike than it seems.

What Makes Social Engineering Trashing Possible?

There's a host of guides to trashing for hacker social engineering purposes. Some were published by phone phreaks in 1970s zines, others by computer hackers in 1980s-era computer bulletin board services. Still others were in online forums (some still in existence to this day). These days, there are even professional sources, entire books dedicated to the topic. All of these sources will tell you that trashing is pretty straightforward: you jump into a dumpster and start looking for potentially useful documents.

But this does not answer a key question: how did we arrive at a point in history when such a practice is even viable, thinkable, and practicable?

There are two key answers. The first is that engineers developed both an infrastructure for and a social habit of wasting in the United States. The infrastructure was developed by Progressive Era sanitary engineers. But that alone is not enough. The mass social engineers of the early- to mid-twentieth century also helped established wasting as a necessary part of the emerging consumer-driven economy of America.

The second answer is more philosophical: the production of trash is in part an act of forgetting. We forget all the time, throwing away old items we no longer want and not thinking about them again. But this is not only about consumerism; it is a fundamental part of knowledge creation. We work our way through ideas, discard old ones, and keep knowledge we find useful. The detritus of such knowledge production is thrown out and forgotten—until a social engineer digs it up and reminds us of it.

Making a Trash Society: Sanitary Engineering and Mass Social Engineering

Underpinning contemporary wasting processes is an invisible infrastructure, best symbolized by the development of landfills across the United States. These systems were developed around the turn of the twentieth century by an emerging professionalized class of engineers who referred to themselves as "sanitary engineers."[11] Their success was deeply transformational: American cities moved from having "garbage [as] an ugly canker staring everyone in the face" in the late nineteenth century to garbage as a "nearly invisible canker, so easy to forget as it swelled beneath the surface" in landfills.[12] Trash was moved from city streets to hidden sites.

The sanitary engineers' success in developing these infrastructures led to some very sweeping proclamations about the power of engineering to change society. Some of the early twentieth-century sanitary engineers

> perceived a higher calling: sanitary engineers had to transcend their training and seek larger roles as community leaders in philanthropic and political capacities, especially as members of civic commissions, as municipal administrators, and even as officeholders. As Ellen Swallow Richards, instructor in sanitary chemistry at MIT, a pioneer in the field of ecology, and a leader in the home-economics movement, stated, "The sanitary engineer has a treble duty for the next few years of civil awakening. Having the knowledge, he [sic] must be a 'leader' in developing works and plants for state and municipal improvement, at the same time he is an 'expert' in their employ. But he must be more; as a health officer he must be a

'teacher' of the people to show them why all these things are to be."[13]

Here we see another example of the seductive power of engineering: despite their claims to political neutrality, some engineers saw their successes in transforming society with sewers or landfills as a warrant for expanding the engineering mentality into other domains, including political administration.

And of course, the engineers who took these stirrings of political power and ramped them up are the mass social engineers who emerged during the Progressive Era. They also played a role in developing a wasting society, especially from the 1920s through the 1950s. The development of invisible infrastructures of waste, along with increased productivity due to Fordist manufacturing processes, enabled a consumerist ideology in which fashionably, technologically, or luxuriously "new" objects must constantly replace the "old," which are subsequently forgotten about. The mass social engineers taught us that wasting is essential to the economy.

As the journalist and social critic Vance Packard argued in 1960 in his book *The Waste Makers*, what kept mid-century marketers up at night was the "the specter of glut for the products they are already endeavoring to sell."[14] The 1950s were especially marked by "bulging inventories of goods" that had to be sold.[15] And to sell those goods, Americans had to be convinced to throw away the perfectly good items they already owned. "Failure to waste was the enemy" of the mid-century marketers.[16]

An obvious way to prevent the failure to waste was *planned obsolescence* of big-ticket items, such as cars and appliances. As Packard argued, planned obsolescence could be engineered into the products themselves, with parts that wear out just after the last payment was made on the credit plan for the item. More effective, however, was *psychological obsolescence*, or designing a product in such a way as to "strip it of its desirability even though it continues to function dutifully."[17] Borrowed from the world of fashion, such psychological obsolescence was applied to automobiles in the form of model years, fins, different colors, or modifications to the grills. Notably, Packard had diagnosed this practice in 1957 in another major book of his, *The Hidden Persuaders*. Packard connected the advent of psychological obsolescence to "consent engineering," calling figures such as Edward Bernays "merchants of discontent" who taught Americans to toss out the old in favor of the new.[18]

While psychological or planned obsolescence is usually used in reference to big-ticket items like cars or washing machines, our interest here is in dumpsters and the more quotidian waste that they contain. Returning to concerns about sanitation, we could highlight the seemingly innocuous disposable paper cup, such as the Dixie Cup, developed in the early twentieth century. Reportedly, the consent engineers Bernays and Fleischman helped advertise Dixie Cups as a more sanitary replacement for washable cups. Their advertisements included images of disease and filth to suggest that the one-use paper cup would save Americans from illness.[19] Rather than reuse a glass cup, Bernays, Fleischman, and Dixie Cups tell us, just use the paper one, throw it out, and forget it.[20]

But a major change that underpins what would become phone phreak and hacker trashing is the advent of credit. The expanse of credit in mid-twentieth-century America meant that "much of the material buried in landfills in recent years was bought with . . . credit cards, leading to the quintessentially American practice of continuing to pay, sometimes for years, for purchases after they become trash."[21] Credit would not only speed up circuits of consumption and wasting; it would also be a system to produce an expanding class of paper waste. After all, to pay for the big-ticket items in the mad dash to stay ahead of psychological obsolescence, Americans would need a growing credit bureaucracy, complete with paperwork, bills, and statements.[22] Such documents contain a wealth of information—who has access to credit, who owes what, and which institutions are offering them credit.[23]

Certainly, such credit documents are regularly filed. However, very often, the same out-with-the-old logic that underpinned wasting of everything from paper cups to automobiles applied to bills and credit statements: it could all go in the trash, out to the dump to be forgotten. Corporate employees might dump such paperwork, and certainly consumers dumped old bills and statements. Credit documents that show the relation between payment and used goods often get buried alongside the trashed products that were paid for with credit. Those credit documents were prized finds when hackers and phreaks went trashing.

The overall result is a transformation of American society, where citizens become consumers, conduits for the flow of commodities to become waste, speeding up the flow with credit. But beyond this wholescale transformation of American political economy into a wasting economy, there is another answer to what makes hacker trashing possible. It's a philosophical answer, one that ties traditional waste directly to our contemporary digitally mediated society.

A Philosophy of Garbage

Garbage is not just the byproduct of consumerism. It's also a result of knowledge production.[24] In John Scanlan's cultural history of garbage, he draws on the philosophy of Immanuel Kant to argue that reason's dark side is comprised of waste. For Scanlan, Kant's conception of reason "rests squarely on the *disposal* of doubt, error, uselessness, and so on."[25] Thus, to produce knowledge, one must work through the "half-formed ideas, unworkable hypotheses and dead ends of the intellectual endeavour."[26] We may think of knowledge production as a clean and orderly process, but of course all knowledge is marked by failure, missteps, mistakes, and accidents. Garbage is the byproduct of the production of knowledge. While accepted knowledge is valued, trashed ideas are reduced to a semiotic jumble. As Scanlan argues, "all attempts to present a unity of knowledge, or an overcoming of error, results in the creation of garbage . . . so it is that knowledge is more generally regimented for particular ends by the purging of the excess."[27]

Similarly, environmental scholar Myra Hird analyzes garbage from a feminist epistemological perspective and argues that garbage is indeterminate, with each landfill a "heterogeneous, unique mix . . . intra-acting with seasons, weather, precipitation, the varying angles of the sun's rays bombarding landfill material and so on."[28] Waste management—the massive infrastructure of moving garbage from place to place, so often to the landfill—is an attempt to determine garbage, to make it knowable and hence controllable, but such determination always escapes waste managers. Indeed, we have almost no information about waste streams, especially when compared to the sheer glut of information we have on supply chains.[29] This is, ontologically, what makes waste waste, and hence in waste we see how our knowledge (the orderly, determinate Reason) is always entangled with the indeterminacy of undifferentiated trash.[30]

Moreover, thanks to the dual innovations of landfill infrastructures and get-the-newest-version consumerism, we are invited to simply *forget* trash altogether. As Hird argues,

> Diligent middle-class western practices of placing garbage on sidewalks to be taken to dumping stations, landfills, incinerators and the like ritualises this forgetting. It is made possible by legislative decision, regulative enforcement, risk models, community accession and engineering practice. As such, landfills make their appearance on and in the landscape as a material enactment of forgetting.[31]

Thus, not only does our economy rely on the refusal to acknowledge refuse, our very epistemologies also overlook trash as we focus on accepted knowledge.

Garbage thus becomes the dark side to all we value, including knowledge. This shadow appears in any locus of knowledge production, from cooking schools (consider all the failed souffles in the trash, not to mention the not-so-proverbial broken eggs used to make omelets) to art schools (think of the bits of rubber that fleck off an eraser) to the academy (imagine the multitude of paper ideas that have been thrown in the trash can). This includes outdated and old ideas and theories.

Combining all these factors—the infrastructure of refuse, the waste-making political economy, and throwing out the trash as a simultaneous act of knowledge-making and forgetting—and we have the conditions of possibility for dumpster diving. While we may believe that discarded knowledge byproducts have no value and allow them to return to the world of indeterminacy and chaos, while we may toss them aside and thus forget about them, trashing is possible because the glut of wasted byproducts can be reconstituted into new forms of knowledge. Such knowledge enables new, manipulative communicative practices. Scraps of paper, a credit receipt, a phone company manual, an old invoice, a printout of a mass email message—any number of heterogeneous bits and pieces of undifferentiated waste can be drawn upon to make deep insights into individuals or organizations.[32] And, as we will argue below, these logics extend into the world of digital media.

The liberal use of the waste system as an endpoint for the byproducts of knowledge production creates a steady and easily located stream of such material. All the hacker or phreak has to do is find a key node in the larger garbage network to target for potential information. This node is the interim point between consumer-side wasting and the landfill: the dumpster.

In the Dumpster

After all, the garbage has to go somewhere before it's picked up for transportation to the landfill: cans by desks, bins under sinks, "herbie-curbies" by the road, or dumpsters behind the office. The dumpster in particular is a holding cell, a place for trash to sit and wait for its fate in the sorting facility, recycling center, or, most likely, the landfill. It is an interim place between short circuits of consumption and the long circuit of landfilling.

It is also a target.

The hacker guides to trashing—also called dumpster diving—found across hacker zines, bulletin boards, and online forums recognize this target. In a lengthy guide in *2600* in 1984, two phone phreaks instruct the would-be trasher to focus on phone company dumpsters. The first step, they suggest, is reconnaissance: When does the trash get picked up? Who is patrolling the parking lot? When it comes time to examine the dumpster itself, the authors are clear: be prepared to jump in. Do not sort in the dumpster; just grab anything that looks promising. Use a car to get to and from the dumpster. Never sort on site. Sort at another location. And above all, avoid cafeteria dumpsters.[33]

This guide—representative of many—encapsulates the value transformation methods of trashing. The reconnaissance stage requires the trashers to locate the precise dumpster in which discarded knowledge may be waiting. Jumping in and grabbing is a physical act of labor. Transportation in a car, where the garbage is hauled away from the dumpster and hence away from the landfill, reverses the wasting process by which the office workers believed they were ridding themselves of useless knowledge. And sorting is key: to use Hird's language, it is determining the indeterminate, remembering what was forgotten, again through labor (sorting, categorizing, thinking). What information is potentially useful for social engineering? What material is best thrown back?

As for what materials may be determined to be of value, as a computer hacker writes in *TAP* in 1982,

> Garbage can provide an excellent source of information ... dialups, passwords, computer logins, etc. Try going through your local

ESS dumpster late at night and see what's there. Or perhaps a stock market place. Or bank (be careful, though). I think you'll be surprised at what you'll get (besides fleas and the plague)![34]

Also in *TAP*, an expert credit card fraudster recommends finding carbon copies:

> To credit card things . . . you search the dumpsters outside of any store that takes credit cards, look for the pieces of carbon paper that they use when running off a credit card sale. This is called "trashing."[35]

In her interview with Geraldo Rivera, the phone phreak, hacker, and social engineer Susan Headley drew attention to discarded phone company manuals:

> [Phone company employees] get an update, they throw away the old [manual], just toss it in the bin. So the phone phreaks would go through and clean out the bins. There was a lot of knowledge retrieved that way.[36]

Further, although not specifically aimed at social engineers, John Hoffman's book *The Art & Science of Dumpster Diving* is cited in BBS posts as an excellent guide to the practice, particularly his chapter "Information Diving." In that chapter, he argues that

> Trash can tell you EVERYTHING about a person: what he eats. Where he shops. His income. His hobbies. His doctor and medications. Where he works, plays and stays overnight. And, of course, you might get lucky and find something really good: A letter from his escaped felon brother. A note from a mistress. Old documents alluding to some long-buried scandal.[37]

While these guides date back to the early 1980s through the 1990s, trashing in dumpsters is still going strong today. There are now guides from professional social engineers on the topic. Social engineering specialist Chris Hadnagy's 2010 book *Social Engineering: The Art of Human Hacking*, includes a section on dumpster diving, including basic legal and health advice, as part of penetration testing for professionalized social engineers. Sharon Conheady's 2014 book *Social Engineering in IT Security* does, as well.[38] Video guides to the practice abound on YouTube. Perhaps the most elaborate guide to dumpster diving for hacker social engineers is Johnny Long's 2008 book *No Tech Hacking*, which dedicates its entire first chapter to the practice. Thus, despite years of the information security professionals recommending paper shredders to corporations, trashing remains an attractive option for the initial research phase of social engineering.

The Digital Dumpster

Of course, some may argue we don't use paper anymore, that our information has been digitized, which surely does not have the same material footprint as physical trash. We even see articulations of digital media with "going green": "Avoid getting paper mail," our banks and credit card companies urge us, "and switch to eBilling!" These arguments are false for many reasons: not only do we still use a lot of paper, digital media is hardly ecologically guilt-free, relying as it does on a vast, resource-intensive material infrastructure to function.[39] A number of scholars, notably in media studies and legal studies, have begun to consider the vast networks of data not as a clean, hygienic space, but instead as a sort of digital dumping ground.[40] And this globe-spanning digital infrastructure of data—including abandoned, ostensibly deleted, and "outdated" data, arguably a form of data trash—has not gone unnoticed by social engineers, who have updated their techniques accordingly.

Digital dumpster diving's conditions of possibility are quite similar to its traditional counterpart. The same churn that produces the vast amount of waste that goes to landfills—dispose in order to forget, throw away in order to consume—drives contemporary digital media practices. Media studies scholar Mél Hogan calls this the "digital dumpster." For Hogan, the digital dumpster arises from a form of digital hoarding made possible by increased storage capacities:

> We encourage hoarding without conscience, without consequence, without affective insights into what we are doing or who might be affected. Old emails, social media streams, texts, and increasingly, the Internet of things, continue to exist on servers that hoard our data for us, often as a business model for "free" services.[41]

This digital hoarding is a byproduct of the larger move to "self-curate," to construct the self out of bits and pieces of digital media. Today's contemporary digital media companies are direct descendants of the mid-twentieth-century marketers that journalist and cultural critic Vance Packard called "the waste makers."[42] Corporate social media emphasize the new and the trending. Take another profile picture. Update your status. Tell us what's on your mind. These are your trending topics. Join this new service. Download this latest app. All of these actions are marketed as a means of self-expression and authenticity in the digital age. In this constant churn, as new updates appear, older digital artifacts get pushed down the timeline, seemingly off the platform. Or, we leave data behind as we leave an out-of-date platform for the hot new one.

This process is a digital media iteration of the same processes of self-production qua wasting that Hird identifies in her work on garbage. "Landfills swell with things we once wanted and now do not want, once valued and no longer value," Hird argues. "What remains after our disgorgement is what we (want to) consider our real self."[43] Likewise, self-curation practices: what is left over after we discard poorly framed selfies from Instagram, delete a tweet from Twitter, or revoke a "like" on Facebook? What remains after we delete images, text, or videos off our phones or computers? We rely on such digital disgorgement to produce our real selves.

And in doing so, we leave a shadow of digital waste in the vast digital dumpster. The discourse of Big Data is entirely predicated on "recycling" the byproducts of our age, a waste product euphemistically referred to as "digital traces."[44] These byproducts are "the digital exhaust of web searches, credit card payments, and mobiles pinging the nearest phone mast."[45] Of course, we know that the digital media corporations store the material we upload and want to share. What is less talked about is the fact that they also include data we *discard*. For example, Facebook records status updates and text entry that the end user ultimately does not post.[46] Twitter maintains deleted direct messages.[47] Google Docs keeps every version of a document—including every deletion, every mistake, every revision.[48] Corporate social media's efforts to collect "deleted" data should alert us to the value of digital trash. Like any knowledge production process, self-curation *necessarily* produces waste: the selfie that wasn't in focus, the deleted status update, the revoked friend request, the email draft abandoned, the files removed from the hard drive. While we may ignore these media mistakes and discount them as not part of our self-curation practices, like regular trash, they are eloquent witnesses about who we are.[49] Moreover, often digital data are in fact stored in digital versions of dumpsters: "trash" or "recycling" folders on hard drives or search engine caches. These data are an attractive target for contemporary social engineers who engage in digital dumpster diving.

Trashing in the Digital Dumpster

Phone phreak and hacker social engineers recognized digital dumpster diving possibilities quite a long time ago. In fact, in her testimony to the United States Senate in 1983, Susan "Thunder" Headley explicitly linked traditional trashing with computer-based, digital trashing. She testified that hackers would look through regular garbage but also "free disk space [on computers] because when something is deleted it usually isn't zeroed [that is, completely deleted]."[50] Likewise, in an anonymous article in *2600* in 1982, an author noted,

> Garbage picking is the art of finding things that someone else has thrown away. Hackers on time-sharing systems are long familiar with the technique of asking the operating system for some memory or mass storage space that has not yet been zeroed out, and then dumping out whatever was in there to the screen or printer. Things like password files and system programs are always updated or backed up from time to time, and that's when a "garbage copy" will be created.[51]

This technique is called making a "core dump." Much like regular trash, when we delete files off computers, we may be encouraged to think of the data as truly destroyed, but often it can be recovered.[52] Deleted data thus becomes a ripe target for those seeking to build profiles, map organizations, gather passwords, or find software vulnerabilities to exploit.

More recently, digital trashing has become highly networked. As the internet became more and more popular, hackers "armed only with a search engine and a bit of ingenuity" created "Google hacks" or "Google dorks," which are highly targeted searches designed to find publicly accessible data.[53] These data include temporary caches, such as .trash, recycler, or $Recycle.bin folders on web server operating systems.[54] These searches for digital trash can be limited to targeted URLs. For example, if someone were looking for temporary files hosted on the University of Utah's website, they would limit the search to the institution's domain (www.utah.edu) and add in the Google Dork to find temporary files.

We also cannot overlook "data dumps," the posting of previously restricted information gleaned after corporate or government databases are breached (often, we should note, using hacker social engineering techniques). Whether obtained by hacktivists, leak sites meant for whistleblowers, or simply for bragging rights, massive amounts of previously restricted information have been "dumped" on the internet. As security consultant Michael Bazzell advises listeners on his podcast, "If you are not using breach data as part of your online investigations, I believe you are really missing out."[55] A notorious example was the exposure of millions of email addresses from Ashley Madison, a website marketed to married people seeking affairs. *Wired* magazine reported that, in addition to email addresses, the 10-gigabyte data dump "also contains internal documents from Ashley Madison including employee credentials, charts, contracts, and sales documents."[56] As Dave Kennedy, a longtime participant in the *Social-Engineer.org Podcast* and author of the Social-Engineer Toolkit software suite, writes with his colleagues, "This dump appears to be legit. Very, very legit."[57] Here, he's not only verifying that the data included in the dump are real; he's verifying that anyone can dive into the dump and gather an incredible amount of personal and corporate information on potential targets.

Finally, we should not forget digital waste in what we might call the old-school, physical dumpster. Given our contemporary version of planned obsolescence—that is, the idea that a digital device is obsolete after about 18 months—dumpsters all over the world are brimming with discarded phones, tablets, hard drives, and USB drives. One security researcher went through some old digital devices he acquired from recyclers and resellers and found that

> The pile of junk turned out to contain 41 Social Security numbers, 50 dates of birth, 611 email accounts, 19 credit card numbers, two passport numbers, and six driver's license numbers. Additionally, more than 200,000 images were contained on the devices and over 3,400 documents. He also extracted nearly 150,000 emails.[58]

All these data were culled from only eighty-five devices.[59] We have no reliable estimates of how many devices are in dumps around the world, but it's fair to guess the number is easily in the millions.

Conclusion: Surveys, Dumpster Diving, and OSINT

Given the vast amount of information available online, it seems far less likely that a contemporary social engineer would need to sully themselves in a physical dumpster. As professional social engineer Sharon Conheady noted with glee during a security conference presentation,

> People publish so much information [in social media] that they don't realize could be used in a social engineering attack. And what I like about it is: I don't have to do so much dumpster diving anymore 'cause the information is available online! And I hate dumpster diving—it's so dirty![60]

Moreover, even poring through digital data dumps is not often called "digital dumpster diving." Contemporary hacker social engineers—especially professionalized security consultants like Conheady, who gets paid to use social engineering to test organizational security—use a cleaner term: OSINT. Appropriated from the military, OSINT refers to "open-source intelligence" gathering; that is, "the repurposing of public records for intelligence and investigations, including social media content not protected by privacy settings."[61] Tools of the trade include Google Dorks, specialized search engines, social media APIs, and consumer databases.[62] OSINT is not just the domain of hacker social engineering. It has become a ubiquitous term in the digital age. Journalists, for example, are starting to use the term to characterize the work of rooting through "satellite imagery, social media, databases of wind, weather, and vessel movement" to tell stories.[63] OSINT not only refers to a seemingly different approach to gathering data than the older dumpster diving method, its militaristic acronym gives it an air of respectability among the security-minded—certainly more so than "trashing" or "dumpster diving."

However, for all its slick, software-driven aspects, OSINT is a genealogical descendant of the older political economy and epistemology of garbage and wasting. OSINT "enables the recasting of otherwise ephemeral social interactions into raw data . . . to be analysed."[64] Like the paper tossed in a trash can, what was forgotten or seen as transitory by one party becomes valuable data to another. The picture posted late one night a year ago, the social media comment made in passing and then deleted five years prior, or even a connection made to a questionable website a decade ago may seem lost in the digital dumpster, but the digital dumpster divers of OSINT research may recover them. We see vestiges of the older phreak trashing practices in contemporary OSINT tool names, especially dnsdumpster.com, a site that returns domain name system information on web servers. There is a direct line from the waste that the early twentieth-century sanitary engineers had to deal with in the wake of the glut of Fordist production to what media studies scholar Mark Andrejevic calls the "infoglut" of the digital age.[65] In order to do their work, social engineers must pore through glut.

Moreover, the epistemological dimensions of trashing, including the recovery of forgotten knowledge, helps us reconsider the research practices of mass social engineering. A 1942 article on "advertising engineering" argues that efficient mass social engineering requires intense study of consumption practices in peoples' homes. The article explicitly calls for examination of trash. This is because trash tells the truth while people forget: "Nothing is left to the housewife's memory. She saves her empty packages."[66] In other words, they ask their target families to save their garbage, lest they forget what they consumed.

The social engineering practice of trashing betrays both the conceit that knowledge production is a clean process with no byproducts, as well as the false anti-materialism implied by the willful ignorance of the infrastructure of the landfill. We might forget what's out there—the waste online or in the dumpster. But when the document is recovered, the information extracted, and the organization, demographic, or person profiled through the trash, the social engineer exposes our collective ignorance about the informational potentials of our waste. As hacker social engineer Johnny Long caustically puts it, "Go on thinking your personal or corporate secrets aren't sitting exposed in a dumpster somewhere, waiting for a no-tech hacker to snatch them up."[67]

Armed with garbage-begotten knowledge about us, the social engineer can move to the next phase, constructing a false, yet plausible and recognizable, identity or role to play: the pretext.

4
Pretexting: Recognizing the Mitnick Mythology

> The goal of the social engineer is to get you to make a decision without thinking. . . . I need to state one very important point: social engineering is not politically correct. . . . [social engineering] takes advantage of the fact that gender bias, racial bias, age bias, and status bias (as well as combinations of those biases) exist.
>
> —Chris Hadnagy[1]

> We all have stereotypes which minimize not only our thinking habits but also the ordinary routine of life.
>
> —Edward Bernays[2]

For several years, the *Social-Engineer.org Podcast*, a flagship podcast in the field of hacker social engineering, used a song called "Trust Me" as its theme song. Written by the nerdcore hip-hop duo Dual Core, "Trust Me" includes a boast about the social engineer's ability to take on any role:

> I can be anyone, anywhere I aim
> I just need the mindset, a story, and a name.[3]

Dual Core's lyrics bring to mind the clichés of spy movies, where spies quickly don disguises that get them into the most secure areas. The song's subsequent lyrics continue this theme, with rapper Int80 exploring a host of roles he could play: clipboard-carrying foreman, customer service worker, corporate executive, tech support.

Hacker social engineers have a word for this: pretexting. Put simply, a pretext is the role a hacker social engineer will play when they are engaging with the target. Obviously, a hacker can't call up an organization and say, "Hello! I am a hacker. May I please have a password to your network?" Instead, the hacker social engineer practices deception by playing a role with a seemingly legitimate need for sensitive information. Logically, pretexting follows trashing quite nicely: trashing provides a wealth of information that can inform the roles a social engineer can play.

Here, we take up this hacker social engineer term with several goals. We'll explore the interpersonal dynamics of hacker social engineering, and then we'll use the concept as a lens to look backward at similar mass social engineering practices. Just as in Dual Core's "Trust Me," we'll see a range of pretexts: construction workers, representatives of charitable organizations, citizens' councils, and medical professionals. This basic analysis will appear to reinforce Dual Core's observation that social engineers can be anyone, anywhere they aim.

However, the heart of our analysis is not about the social engineer's wild abilities to take on any role they please. Instead, we focus on the social structures that constrain as well as inform social engineering pretexts. To explore such social structures, the bulk of our chapter engages with perhaps the most famous social engineer of all, Kevin Mitnick. Mitnick started as a phone phreak, became a felon after being convicted for hacking crimes, then later evolved to be the respected security consultant he is to this day. Stories about Mitnick's exploits abound, giving us the impression of a man with mythical pretexting abilities. If anyone can be anyone, anywhere he aims, it might be Mitnick.

And yet, as we argue, Mitnick's individual genius is grossly overstated. Mitnick's pretexts were reliant on social structures. The lesson we learn from Mitnick's case is that it's not the virtuoso abilities of the social engineer that enables a pretext to succeed. Ultimately, pretexts succeed thanks in large part to the pernicious persistence of reductive stereotypes. We will see the power of reductive stereotypes not only in hacker social engineering, but also in mass social engineering and later in masspersonal social engineering.

Hacker Pretexting

The pretext is an exceptionally important aspect of a hacker social engineer's attack; it is not to be left to chance. It must be methodically planned out in advance, and the literature from professional social engineers reinforces this idea. The social engineer needs what security consultant Sharon Conheady's book *Social Engineering in IT Security* refers to as "the backstory that explains who you are and why you need the information you are requesting." The pretext "even influences your attitude while executing the test."[4] Conheady's colleague, Chris Hadnagy, invokes a pseudo-method acting description of the process:

> Pretexting involves not just coming up with the storyline but also developing the way your persona would look, act, talk, walk; deciding what tools and knowledge they would have; and then mastering the entire package so when you approach the target, you are that person, and not simply playing a character.[5]

Such attention to the pretext is needed because hacker social engineers directly engage with their targets in interpersonal communication: on the phone, over email or text, or even in person. The role, then, must be carefully prepared before it is put to trial during an engagement.

These roles can vary based on what the social engineer learns during the trashing phase. Does the target donate to charities? Perhaps I should email a malicious PDF of a charitable society's flier. Does the target contract with a particular cafeteria vendor? I can mimic their uniforms and get into the organization that way. Does the target use a particular type of computer? Perhaps calling up and saying that I'm with tech support and I need to update their operating system can work.

During these pretexts, social engineers use the "stuff of authenticity": material artifacts that align them with the organization they seek to infiltrate.[6] Common artifacts are uniforms, ID badges, lanyards, business cards, hardhats, lunchboxes, toolboxes, business attire, phones, or (if one is trying to tailgate through a door) several to-go cups of coffee.[7] Consider this social engineer's guide to visual (and, to a lesser extent, olfactory) authenticity during a pretext as a construction worker:

> A common mistake is to purchase brand-new high visibility vests and hardhats, which stand out as somewhat unusual. How many workmen [sic] are seen with a pristine outfit? When the [social engineers] eventually built up the wardrobe . . . , they swapped brand-new high visibility vests and hats for used ones. Unfortunately, the used items stank of diesel, but at least they looked authentic, because they were.[8]

Here, used and thrown-out clothing can help with playing the role.

Even in the case of pretexting over the phone or over digital channels, all the right details need to be in place. If a hacker social engineer is making a telephone call and playing the role of an office worker, they'll probably want to play office sounds in the background; a phone center employee pretext may call for a different soundscape (both easily available via YouTube videos).[9] If they're using email, a company logo or spoofed website may be in order.

And all of this is not to mention using the proper tone during the phone call or in written correspondence.

Mass Social Engineering Pretexts

While the interpersonal hacker social engineers have spent a great deal of time focusing on constructing roles and scenarios, they weren't the first social engineers to do so. As pioneering public relations practitioners in the 1920s, mass social engineers Edward Bernays and Doris Fleischman regularly created "circumstances which will modify" the habits and customs of targeted groups.[10] To do so, these mass social engineers used pretexts just as hacker social engineers would do. However, unlike hacker social engineers, mass social engineers did not adopt the roles themselves. Instead, their created circumstances were populated with surrogates who played the roles. Bernays wrote about this explicitly:

> the public relations counsel . . . may enlist the interest of an individual or an organization in his client's point of view. . . . That individual organization may then propagandize [the client's issue] through its own channels because it is interested in it. In such a case, *the point of origin then becomes that individual or organization.* The public relations counsel, having made the link between the interest of his client and the interest of the third party, no longer need figure in the resulting expression to the public.[11]

If there were no existing groups or individuals willing to propagandize on behalf of Bernays and Fleischman's clients, they would simply borrow from the playbook of the social reformers of their time and create

> seemingly independent organisations which profess to support concerns of the common good: the Committee for the Study and Promotion of the Sanitary Dispensing of Foods and Drink; the Radio Institute of the Audible Arts; the Temperature Research Foundation; [and] the Middle America Information Bureau.[12]

While these organizations imitated the early twentieth-century progressive social reformers' use of citizen committees, their purpose was not for self-governance. Instead, they were always tied to a specific public relations campaign Bernays and Fleischman were hired to conduct. For example, the American Council for Wider Reading was formed while they were doing public relations for a book publisher.[13] The Committee for the Study and Promotion of the Sanitary Dispensing of Foods and Drink was reportedly created to promote Dixie's disposable paper cups as more sanitary than reusable, washable cups.[14]

One of their efforts, a wildly successful idea Bernays and Fleischman pitched to Proctor & Gamble in the early 1920s, deserves more mention: soap carving competitions. These were created to support P&G's goal of selling more bar soap.[15] Starting in 1925, P&G began sponsoring contests for Americans to create art carved out of Ivory soap. But rather than appear to be a "string-pulling mastermind," P&G established a "National Soap Sculpture Committee" as a pretext:

> This sounded very official, indeed, and its New York City address only punctuated the authenticity of the committee's art world credibility. All aspects of the contest were handled publicly under this name. The committee published every contest announcement and exhibition catalog, and it was responsible for an informative series of books by soap carvers on the how-tos and wherefores of their chosen art form.[16]

P&G was never mentioned as the "contest's sole originator and coordinator (thus 'donating' money only to its own PR programs)."[17] As Jennifer Jane Marshall writes in her article "Procter & Gamble's Depression-Era Soap-Carving Contests," "This way P&G was able to avoid the appearance of impropriety, an important measure of decorum in an era when advertising gimmicks—and exasperation with them—ran high."[18] Indeed, even the name of the contest was a front: "The National Soap Sculpture Competition in White Soap" didn't mention Ivory by name, but Ivory was the only white soap on the market at the time of the contests.[19]

In other words, the National Soap Sculpture Committee was a pretext, playing the role of sponsors of wholesome art contests but designed by the mass social engineers Bernays and Fleischman to help P&G sell soap. Sales of Ivory did indeed rise—even during the Great Depression.

But perhaps the most infamous pretexts concocted by Bernays and Fleischman came from their late 1920s work for American Tobacco, a company seeking to increase the number of women smokers. This work relied heavily on pretexting. They linked smoking to thinness and health by enlisting experts to condemn sugar and praise cigarette smoking. Bernays and Fleischman solicited a physician who obliged them with this medical advice: instead of having a sugary dessert,

> the correct way to finish a meal is with fruit, coffee and a cigarette. The fruit hardens the gums and cleans the teeth; the coffee stimulates the flow of saliva in the mouth and acts as a mouth wash; while finally the cigarette disinfects the mouth and soothes the nerves.[20]

Bernays and Fleischman shared this quote with journalists in order to get news coverage of the physician's recommendation. To bolster the claim that cigarettes are healthier than sundaes, Bernays and Fleischman recruited dance school instructors, photographers, and dance troupes to praise being thin, cutting out sweets, and smoking cigarettes.[21] Again, they shared these sentiments with reporters, creating the appearance of a nationwide consensus against sugar (and, incidentally, in favor of cigarettes). This way, their client, American Tobacco, never had to make direct claims; these proxies did.

In addition, Fleischman and Bernays linked cigarette smoking to feminist emancipation, recruiting women to march in the 1929

Easter Parade, carrying lit cigarettes as "torches of freedom" to protest a taboo against women smoking in public.[22] The goal was to have this seemingly subversive act covered in the news. The women did not reveal that they were recruited to march by a PR firm hired by American Tobacco.[23]

Finally, in order to encourage more women to smoke American Tobacco's Lucky Strike cigarettes—which came in distinctive green and red packaging—Bernays and Fleischman concocted a "Green Ball" at the Waldorf Astoria hotel in New York City. While American Tobacco sponsored the ball, Bernays and Fleischman hid that fact behind a front of New York socialites who took credit. The ball required everyone to wear green, and it was so successful in its "propaganda efforts," recalls Fleischman, "that the country was swept by a demand for green costumes and accessories."[24] And again, American Tobacco was never openly named.

These practices were not atypical for mass social engineers—indeed, histories of public relations are rife with documentation of "front groups" who stood in for powerful organizations—but Bernays and Fleischman perfected the form.[25] The use of pretexts is now a common public relations tactic.[26] Today, while multiple professional public relations groups urge their members not to use this "third party technique," a wide range of interests, from pharmaceutical companies to governments, employ mass social engineers who create these pretexts with few to no consequences for such deceptions.[27]

To work, such pretexts had to be believable, and such belief had to start with the mass social engineers themselves. "Whatever cause they serve or goods they sell, effective propagandists must believe in it—or at least momentarily believe they believe in it."[28] Like the recommendation that the hacker social engineer must totally inhabit the pretext they create, the mass social engineer must believe that their work is important, that the cause they champion is beneficial to all of humanity.

The "World's Most Famous Hacker"

This accounting of hacker and mass social engineering pretexts might give the impression that social engineers can take on just about any pretext imaginable. The pretexts used by Bernays and Fleischman are particularly legendary—many commentators present their work on behalf of American Tobacco as being the key reason women increasingly smoked cigarettes in the twentieth century. And our contemporary vision of hacker social engineers is of devious geniuses who can take on a wide range of roles.

But there is one figure in social engineering history who appears to be the master pretexter: Kevin Mitnick, the self-proclaimed "world's most famous hacker."

More than just about anyone in the past century, the most common name associated with social engineering—mass, interpersonal, or otherwise—is Kevin Mitnick. Almost every time we mentioned this social engineering book project to people, they would say, "Oh, you mean like Kevin Mitnick." The FBI manhunt in the mid-1990s, his arrest in 1995 for computer hacking, his conviction, and his subsequent writing, speaking, and security consulting careers have all contributed to Mitnick's self-proclaimed status as the world's most famous hacker. And his hacking technique of choice is not to break into networks through computer-aided techniques, but instead to use a pretext and simply call people up on the phone and ask for access. Mitnick is not just the most famous hacker; he's the most famous social engineer.

Media coverage of Mitnick's criminal career and subsequent reformation and legitimation have not challenged Mitnick's self-mythology.[29] Recent headlines about Mitnick refer to him as "legendary," a "master hacker" who can "access your system in less than an hour."[30] Mitnick himself does little to discourage such praise, bragging to journalists that he is a "rock star in the hacker community" and that "I'm very good at what I want to do."[31] And he reinforces his legendary status with his books, especially *The Art of Deception* and *Ghost in the Wires*. In those books, he relives his social engineering pretexts: on one telephone call, he's a customer. The next, he's a police officer. Next, he's the CEO. And in the end, businesses are infiltrated, information purloined, and Mitnick's status grows. Above all, reading through the news coverage and his books, we learn that the rest of us, we hapless humans, are the "weak link" of computer security, to repeat one of Mitnick's favorite phrases.[32]

We could easily contribute to the Mitnick mythology by repeating stories of his exploits, using them to illustrate hacker social engineering pretexts in action, but ours is a different approach. Emphasizing Mitnick's exploits obscures more than it reveals. Instead of focusing on the virtuoso social engineering skills of Mitnick, we want to focus on the structures that enabled him to succeed. We will focus on how his use of pretexts is aided and abetted by social structures of recognition.

Theories of Identity Play

Social engineers' ability to inhabit roles invites us to explore postmodern theories that hold that identity is entirely malleable. Such ideas were especially powerful in the 1990s as the internet gained popularity and people performed a range of identities in online chats, games, and early social media. Recently, however, an increasing number of scholars across a range of fields are starting to shift their analysis away from the individual-focused performance of identity to the larger social structures that both enable and constrain such performances.

Queer theorists considering trans identities are a good guide here. Synthesizing the postmodern, identity-play theories of identity with sociology's interest in social structures and power relations, Raewyn Connell and Carla Pfeffer have argued that "passing," where, for example, a transgender woman is deemed "successful" insofar as she passes as a cisgender woman, places far too much emphasis on the particulars of the woman's performance.[33] Instead, both argue for the concept of "recognition," focusing on how such performances are accepted or denied by those with the privilege to do so. For Connell, recognition is a relational perspective. While a focus on passing may help us to understand the performative aspects of social categories (such as male/female, queer/straight, or Black/white), recognition helps us understand the structures in which such performances are legible.[34] Building on this, Pfeffer argues that,

> rather than focusing on transgender social actors' accomplishment of normative gender through "passing," sociologists might focus, instead, on the interactional processes whereby all social actors serve as arbiters of the gender order as they recognize or reject others as "belonging" to (or rightful members of) particular gender and sexual identity categories and groups.[35]

Such a relational, interactionist conception of how identities are enmeshed in social relations can be seen in the enigmatic philosophy of Michel Serres, whose book *The Parasite* puts forward a triangulating theory of three-part relations and communication.[36] For Serres, communication is not simply a matter of one entity sending messages to another. There has to be a third term, a channel between the two entities. One special type of a third term is the "blank" or "joker," a concept Serres draws from card and domino games:

> The joker is a card in a game that serves to alter the direction of play. It interrupts the game and makes a new set of moves possible. Likewise, the white or blank domino can change the fortunes of a player because it can be played to link sequences of dominoes that are otherwise incommensurable.[37]

Hence, while it is wild, the joker or blank takes on value only in relation to other elements in the game. Its presence is part of the game. Though its special capacity to change its role depending on its context allows for a more dynamic game, it must be played in relation to existing cards.

Bringing these perspectives together, a joker or blank domino can only function within a game if the players both agree upon the rules of the game and recognize its relationship to other cards or dominoes. Likewise, identity performances—including pretexts—work insofar as they are recognized and legitimated by other members of a community who operate under often unspoken social norms and rules. To focus solely on passing ignores the legitimating processes of the other social arbiters who implicitly or explicitly judge the performance and subsequently include and support or exclude and diminish the person attempting to pass. Likewise, focusing solely on the wildness of the wild card, joker, or blank domino is to overstate its capacity to change the game. The joker or blank "certainly adds disruption to social order, but we should not lose sight of the fact that it can also save rather than destroy order."[38]

Thus, role-playing is not simply a matter of taking on a role; it requires others to recognize and accept the role-player. Social structures can constrain, but they allow for creative identity performances within those constraints. As organizational studies scholars argue, "'identity' is a matter of claims, not character; persona, not personality; and presentation, not self. . . . 'Identity' is discursively fashioned by *both* the observers and the observed."[39]

Turning back to Kevin Mitnick, one way to tell his history of social engineering is to focus on his virtuoso pretexts, and indeed many commentators do precisely this. They imply that his interpersonal communication skills are such that he could take on any pretext. This mythology arguably aids Mitnick today, since he claims the title "the world's most famous hacker" and owns a security consultancy boasting of having a "100% success rate" in penetration testing.[40]

However, another way of understanding Mitnick—and by extension, social engineering pretexts as a whole—is to consider his capacity to act as a social joker, a blank that is capable of taking on roles thanks not just to his own pretexting skills but also to a wide range of social structures of recognition. Within these metaphorical rules of the game, Mitnick's varying roles offer a chance to preserve existing orders as much as they disrupt them. This approach draws attention to the social webs that make pretexting possible, showing how pretexting can help maintain—rather than subvert—those social webs. When we read the reporting on Mitnick, as well as his own writings, from this perspective, we start to consider how social structures of social capital, transnational corporate organizational dynamics, and racial stereotypes (particularly, juxtapositions of Asianness against whiteness) help create the conditions that allow Mitnick's pretexts to become recognizable.

Recognizing Mitnick's Pretexting Successes: Structural Factors

Social Capital

Perhaps the most commonly told story of how Kevin Mitnick awakened to the promises of social engineering is his figuring out how to ride public buses for free throughout the San Fernando Valley of California as a child in the 1970s. He noticed that the bus system used paper transfer slips validated with a special paper punch, and he surmised he could ride the buses for free if he could punch his own transfers. In Mitnick's memoir, he tells of what may have been his first pretext, playing the role of a student in need of supplies:

> I walked to the front of the bus and sat down in the closest seat to the driver. When he stopped at a light, I said, "I'm working on a school project and I need to punch interesting shapes on pieces of cardboard. The punch you use on the transfers would be great for me. Is there someplace I can buy one?"[41]

The pretext worked: the bus driver told Mitnick where the paper punches were sold. Later, Mitnick did some trashing to get books of unused transfers from a dumpster behind a bus depot. From that point on, he rode for free "everywhere the bus system covered—Los Angeles County, Riverside County, San Bernardino County."[42]

Thus, Mitnick's career starts with transit fraud. This is a crime that many researchers have noted gets prosecuted quite unevenly, with disparities playing out along racial lines: Black and brown citizens bear the brunt of transit cop attention.[43] But, as the white Mitnick recalls,

> Did I get into trouble for Dumpster-diving for those bus transfers and riding for free? . . . [N]o. My mom thought it was clever, my dad thought it showed initiative, and bus drivers who knew I was punching my own transfers thought it was a big laugh. It was as though everyone who knew what I was up to was giving me attaboys.[44]

In other words, rather than having what Black and brown parents call "the talk"—that is, the warning that American society will viciously prosecute or execute young Black and brown men for even minor transgressions—the Mitnick family either turned a blind eye to young Kevin's pretexts and transit fraud, or they praised him for beating the system.[45]

From this moment on, a pattern emerges: Mitnick goes forth and revels in the identity-play possibilities of telecommunications, using a range of pretexts to gain access to restricted information and software, and, when he is caught, retreats to his family, particularly the loving women in his life. His mother, for example, allows him to use a pretext as an apartment building manager to con GTE out of phone service after the phone company shut down their connection; later, she laughs off a visit from the FBI. "What harm could a boy come to just from playing with a computer at home?" his mother asked. As Mitnick admits, "she had no concept of what I was up to."[46] His mother, grandmother, and aunt provide money for suits, tuition, attorney's fees, and bail. Later, Mitnick's wife Bonnie also supports him (although she does eventually divorce him because he continues hacking). All of them drive him around, including to work, from work, and from various jails and police stations. He moves from social engineering free rides on the bus to bumming free rides from his family.

As professional social engineer Sharon Conheady notes in her book, *Social Engineering in IT Security*, pretexting can be draining. "Your adrenaline pumps so hard that afterward you are completely exhausted." Likewise, on the *Social-Engineer.org Podcast*, episode 120, the panelists, including former FBI behavioralist Robin Dreeke and social engineers Chris Hadnagy and Perry Carpenter, all cautioned would-be social engineers about the emotionally draining aspects of pretexting.[47] These professionals strongly recommend setting aside time to recover from pretexts, whether they be in-person, online, or over the phone (the channel Mitnick preferred). For Mitnick, having social support from his relatives may have allowed him to engage in pretexts knowing that he had some safe space to retreat to and recover from his adrenaline-pumping engagements. He had a support structure allowing him to go forth, "be anyone, anywhere he aims," and yet have somewhere to return to and recover in.

Indeed, the necessity of social support structures for pretexters is highlighted further when we consider the period in the mid-1990s when Mitnick was on the run from the FBI. He cut ties with his family, changed his identity, and moved to new cities. His account of this period in *Ghost in the Wires*, as well as journalist Jonathan Littman's book *The Fugitive Game*, are marked by exhaustion, paranoia, and uncertainty, as Mitnick flees helicopters, is suspicious of anyone sitting in a parked car, and refuses to get close to anyone lest they betray him.

Communication, Organizational Structures, and Transnational Capitalism

Mitnick's pretexts often exploited a simple fact: many organizations are so far-flung that employees don't know each other personally.

Many of Mitnick's targets—NEC, Bank of America, Digital, Pacific Bell, Oakwood Corporate Housing, and TRW, to name a few—were, in the 1990s, regional, national, or even transnational-spanning organizations with branches in many locations. Often, Mitnick would infiltrate these distributed organizational structures with the relative anonymity of telephone calls to slowly but surely gather small pieces of information. As he describes the process, he would gather

> information about the company, including how that department or business unit operates, what its function is, what information the employees have access to, the standard procedure for making requests, whom they routinely get requests from, under what conditions they release the desired information, and the lingo and terminology used in the company.[48]

What Mitnick learns is how communication constitutes these transnational organizations. As communication scholars argue, communication helps define reality, and thus structures interaction, enacts power, and animates hierarchies within organizations.[49] This is especially true in the cases of geographically distributed organizations, where communication across distance facilitates the very existence of organizational units, such as teams.[50] Mitnick demonstrated an intuitive understanding of how communication structures an organization, and much like a manager, he wanted a larger picture of who speaks to whom in organizations, under what conditions, what terms and language they use, and how organizational power is expressed in those communications.

An example from Mitnick's social engineering of Pacific Bell illustrates this further. Based on his previous reconnaissance (whether through trashing, trading for documents, or phone-based inquiries), Mitnick learns the rituals of communication that members of the target corporations engage in. This knowledge includes familiarity with insider lingo. Using this and

> posing as a technician in the field, I called Pacific Bell's Mechanized Loop Assignment Center, or MLAC, also known simply as the Line

> Assignment Office. A lady answered and I said, "Hi. This is Terry out in the field. I need the F1 and the F2 on 310 837–5412 . . ."
>
> "Terry, what's your tech code?" she asked.
>
> I knew she wasn't going to look it up—they never did. Any three-digit number would satisfy, so long as I sounded confident and didn't hesitate.
>
> "Six three seven," I said, picking a number at random . . .[51]

In order for his pretext as a technician to work, the ritual of asking for information (Mitnick's tech code) and giving it must be satisfied, and the rest of the lingo and technical elements are in place. Thanks to this, Mitnick is able to get the information he seeks—an address for a rival hacker—while the Pacific Bell employee believes that she successfully reconstitutes the organization through her interaction with him.

Mitnick's success here hinges far less on his putative skills than on simple, organizational communication routines. As he explains,

> Why was the lady in Line Assignment so willing to answer my questions? Simply because I gave her one right answer and asked the right questions, using the right lingo. So don't go thinking that the Pacific Bell clerk who gave me [the] address was foolish or slow-witted. People in offices ordinarily give others the benefit of the doubt when the request appears to be authentic.[52]

In other words, Mitnick's performance as "Terry out in the field" was not some virtuoso acting, nor was it reliant upon some sort of hypnosis. Simply put, it was recognized by the operator he called as a seemingly normal part of day-to-day communicative structure that constituted the large, regional organization that is Pacific Bell.

After having mapped the organization via its communication network, Mitnick could use the relative anonymity of the phone to insert himself into the organization and capture information as it flowed across the network. He would use a variety of pretexts, ranging from claiming to be a specific member of the organization to claiming to be an outsider, such as a contractor or business partner. But he always had to do so *in relation to others*, making sure his communicative practices aligned with those of his target organization.

"My Hero Is Japboy": Mitnick's Yellowface Minstrelsy

Mitnick also took advantage of existing cultural structures, particularly racial stereotyping. Mitnick's—and by extension, American culture's—peculiar relationship with Japanese people in the 1990s informed one of Mitnick's most notorious pretexts.

As several newspapers and books reported, a key person who helped the FBI catch Mitnick during his fugitive period in the mid-1990s was the Japanese-born American computational physicist and security expert Tsutomu Shimomura. At one point, Mitnick broke into Shimomura's computer. In response, Shimomura aided the FBI in hunting down Mitnick. Thanks to Shimomura's help, the FBI caught Mitnick in his apartment in Raleigh, North Carolina, in 1995.[53]

As a Japanese-born American who gained fame in the 1990s for tracking and catching "the world's most wanted hacker," it is not surprising that the American journalistic coverage of Shimomura played up his ethnic background. Also not surprising—but nonetheless deeply disturbing—was the racism of the coverage. The coverage of Shimomura versus Mitnick took on valences of what American studies scholar Joseph Won calls "yellowface minstrelsy," "the use of Asian martial arts, artists and artifacts by non-ethnic Asians for fun and profit."[54] As Won argues, yellowface minstrelsy parallels older white American appropriative practices of blackface minstrel shows

> in the same way that black dance, music and verbal play summarized black culture in the 19th century, Asian martial arts images today comprise in large part what contemporary consumers of television, film, newspapers and magazines know as "Asian (and Asian American) culture."[55]

A *New York Times* profile of Shimomura was one of the first articles on the security researcher, and it subtly invoked visions of martial arts. Discussing Mitnick's hacking of Shimomura's computer,

> It was as if the thieves, to prove their prowess, had burglarized the locksmith. Which is why Tsutomu Shimomura, the keeper of the keys in this case, is taking the break-in as a personal affront—and why he considers solving the crime a matter of honor.[56]

A "matter of honor" invokes a peculiar American fascination with Asian martial arts movies and, through them, the uneasy relationship between American and Asian cultures—particularly Japan—in the 1980s and 1990s.[57] Framing Shimomura's pursuit of Mitnick in such terms gets amplified in subsequent coverage. A *Rolling Stone* article, for example, presented the contest between Shimomura and Mitnick as a battle of "the samurai and the cyberthief."[58] Such coverage is part of the larger anxiety about the mysterious, technologically adept and wealthy Japanese in 1990s America.[59] Shimomura's abilities with computers and digital networks—two things that many people find inscrutable and difficult to fathom—likely intensified the mystery, resulting in the use of familiar and yet exotic popular culture tropes of martial arts.

Perhaps the most egregious examples of the yellowface minstrelsy framing of Shimomura appears when Mitnick comments on his erstwhile foe, especially in the book *The Fugitive Game*, where journalist Jonathan Littman includes transcripts of his phone calls with then-fugitive Mitnick. Several of their conversations were about Shimomura, who by then was tracking Mitnick. Littman describes Shimomura as a "touted samurai."[60] He discusses a picture of Shimomura in *Newsweek* with Mitnick: "Next to the keyboard," Littman tells Mitnick, "I swear it looks like there is a samurai sword."[61] Mitnick replies, "I'm sure he'd like to chop some people's heads off...."[62] Then, according to Littman, "Mitnick does his best kung fu master imitation. 'You dishonored my family. You will die! I'll meet you . . . and we will fight to the death!'"[63] Here, Mitnick's accent is an instance of yellowface minstrelsy, using martial arts clichés to discuss the "matter of honor." To further fuel the antagonism, Mitnick also mocks Shimomura in an online chat, saying, "my hero is japboy."[64]

However, Mitnick did not limit his use of yellowface accents to his private conversations with Littman. The same yellowface minstrelsy stereotyping that informed coverage of Shimomura also aided Mitnick with one of his pretexts, bringing us to another structural context that social engineers might exploit: racial stereotypes. As he writes in *Ghost in the Wires*, Mitnick wanted to "break into NEC's network and download the source code for all the NEC cell phones used in the United States."[65] To do so, he first used the pretext "Rob in the IT department" to gain access to one computer.[66] But he wanted to get more data, so he called the same NEC employee back, this time with a different pretext altogether. With this second pretext, Mitnick gets, as he claims, "gutsy":

> I was no Rich Little when it came to doing accents, but I was going to try to pass myself off as Takada-san, from NEC Japan's Mobile Radio Division.
> I called [NEC USA employee Jeff Lankford] at his desk. When he picked up the phone, I launched into my act:
> "Misterrrr, ahh, Lahngfor, I Takada-san . . . from Japan." He knew the name and asked how he could help.
> "Misterrr Lahng . . . for—we no find, ahhh, vers'n three ohh five for hotdog uhh project"—using the codename I'd picked up for the NEC P7 source code. "Can you, ahhh, put on mrdbolt [an NEC server]?"
> He assured me that he had Version 3.05 on floppy and could upload it.
> "Ahhh, thank . . . ahhh, thank you, Mr. Jeff . . . I check mrdbolt soon. Bye."[67]

Because this works, Mitnick gives himself a pat on the back: it was an "apparently not-too-pathetic accent."

In invoking a yellowface accent, Mitnick played to the same 1990s American visions of Japan that fueled the journalistic coverage of Shimomura. This was not a nuanced understanding of the many ways the Japanese speak, live, and act, but rather a mode of speaking that satisfied American perceptions of the Japanese. In this framing, Shimomura, who had lived in America since he was six, was no ski bum, but rather a sort of samurai sword-wielding cyber-kung fu warrior on a quest to regain his honor. And Mitnick becomes a rival kung fu master who ultimately fails to best him.[68] Likewise, Mitnick's yellowface accent as Takada-san is a pretext readily recognizable in 1990s America.

In terms of the relational aspects of pretexting, such yellowface minstrelsy benefited Mitnick because it satisfied the recipient of his call, who may himself have drawn on prejudices to judge the legitimacy of Mitnick's call. The accent was apparently "not-too-pathetic," a passing accent, for Lankford. This is a specific case of how stereotypes function during an identity performance:

> Because these structures are so ingrained and because taking advantage of them can offer such concrete advantages, the reproduction of these stereotypes (through passing and to pass) may in the abstract present a challenge to social hierarchies, but in the literal sense also reinforces them.[69]

Whereas we may view a social engineer like Mitnick as subversive and his victims as hapless rubes, taking up the relational aspects of pretexting, we have to question how Takada-san or "Shimomora the Samurai" *reinforce* social hierarchies in the context of the American anxieties about Japan during the 1990s. Who is more recognizably "Japanese"—the staid, long-haired ski bum from Princeton, New Jersey, or the white man fast-talking in a Kwai Chang Caine-esque or Mr. Yunioshi-esque accent? Which accent is more believable: a flat Ohio Valley radio voice or speech punctuated by "ahhs" and dropped consonants? Both accents—the presumably white "Rob in IT" and the faux-Japanese "Takada-san" were ready to hand for Mitnick for his pretexts, and both worked on the same NEC employee, who is invested in helping the corporation function by supporting—and not questioning—his colleagues, the organizational hierarchy, or the dominant racial order.

Social Engineering and Stereotyping

The myth of Kevin Mitnick held that he could do any pretext over the phone and bullshit his way into any information system. Such a myth focuses solely on Mitnick's performances, making it seem as if his abilities are those of a singular genius. We may then extend such mythologies to social engineering more broadly. A social engineer—mass, interpersonal, or masspersonal—can play any role they please and manipulate us. The wild stories of hacker social engineers like Mitnick, the stories from mass social engineers like Edward Bernays and Doris Fleischman, or the specter of Russians manipulating online communication, create the perception of extremely skilled, elite manipulators of targets.

As we have shown, this is not entirely the case. In practice, Mitnick's social engineering pretexts rely heavily on the recognition of others, especially their targets. He operated within webs of relationships and benefited from the recognition of others to pass in the roles he chose to play. In doing so, he was able to be a joker or blank and "[engineer] a kind of difference by intercepting relations."[70] Mitnick on the phone, in the midst of a pretext, *must* be understood in relation to the larger social, organizational, and cultural contexts he was operating in, not independent of them. We've examined the case of Mitnick to show some specific dimensions of this relationality: social capital, organizational structure, and social stereotyping. Here, we want to use the last category, social stereotyping, and return to the broader pretexting practices of mass and interpersonal social engineers we began this chapter with.

One lesson is clear: some bodies and subjectivities enjoy more flexibility than others when it comes to pretexting. The range of options open to white, cisgender males who have a great deal of social capital and speak the dominant language of a region is likely larger than those available to a non-binary person of color who does not have a stable home and speaks with a non-normative accent.[71]

But no matter the embodied existence of the social engineer, their pretexts must be relationally recognized.

For example, in a presentation at a security conference, hacker social engineer and security professional Sharon Conheady takes her audience through a thought experiment: "which . . . pretexts are the most likely to work for me?" She notes she's short and "doesn't look like a hacker"; she presents as female, and her accent is Irish. She then goes through scenarios, asking the audience to judge her likelihood of success. IT department? No, she says, it's awkward; people ask her if she's lost. Teacher? Yes. Cleaner? "Yeah, totally," she says. Waitress? Yes. Construction worker? No. Telecoms engineer? No. CEO? At that point, she states her point: "We play to stereotypes. As much as I hate it, we play to stereotypes."[72] As she notes, her pretext as a CEO over email in a phishing attack might work, but doing so in person would be "pretty brazen." And indeed, given her embodiment and ways of performing her identity, Conheady would have advantages in some of these roles that other people may not enjoy. But she cannot be just anybody. She must be recognized.

That a social engineer has to "play to stereotypes" is not limited to contemporary hacker social engineers. As an analysis of the work of Edward Bernays argues, Bernays

> regards stereotypes as "a great aid to the public relations counsel in his work" because they can be grasped by "the average mind," even though, he acknowledges, they are "not necessarily truthful pictures of what they are supposed to portray." No matter, according to Bernays, [public relations] practitioners can use stereotypes to reach

a public and then add their own ideas to fortify their position and give it "greater carrying power."[73]

For Bernays (and Fleischman), a stereotype is conformity to a particular worldview, consonant with communication theorist Walter Lippmann's formulation of stereotypes as a reductive shorthand.[74] But unlike Lippmann, who is critical of the use of stereotypes, Bernays and Fleischman see in them resources to leverage for mass social engineering. To leverage stereotypes, Bernays suggests paying attention to the "tendency the group has to standardize the habits of individuals and to assign logical reasons for them."[75] Bernays vehemently denies that just any appeal will work for the group being targeted:

> The cause [the public relations counsel] represents must have some group reaction and tradition in common with the public he [sic] is trying to reach. This must exist before they can react sympathetically upon one another.[76]

"Public opinion is the resultant of the interaction between" the public relations counsel and the group mind.[77] And stereotypes help mediate that interaction. As mental shortcuts people use to conform to their groups, stereotypes become powerful tools for social engineering—they are easily recognized. So, returning to Bernays and Fleischman's use of front groups as pretexts, their groups had to be recognized—they had to fit the stereotype—in order to successfully manage the targeted crowds.

Overall, playing to stereotypes is possible, as we suggest, due to the preconceptions of those the social engineer is seeking to engage with. In Conheady's experience, being a petite Irish woman working on servers broke the expectations of others, who asked if she was lost. But being a waitress is easy for them to recognize and fits their notions quite well. For Mitnick, his male voice on the phone was a supple instrument that could play the role of technician, police officer, or manager—or even a cartoonish version of a Japanese engineer, so long as the roles could be recognized. For Bernays and Fleischman, social stereotypes become convenient shorthand in public relations campaigns directed at various groups. These moves just have to be recognizable.

In other words, as Hadnagy and Bernays put it in the epigraphs to this chapter, the social engineer uses stereotypes in order to avoid having the target think. The goal is *recognition*, not cognition.

Conclusion

Pretexting is a staple part of the social engineering approach. While we have largely focused on Mitnick's career in the 1990s, as well as his recollection of it in the 2000s, the dynamics that made Mitnick's pretexts work are reflected across the social engineering literature. Here, we've touched on how the mass social engineers Bernays and Fleischman theorized stereotypes in mass social engineering, and later, we will return to this focus on stereotypes in pretexts in what we're calling masspersonal social engineering. We encourage others to analyze pretexting in terms of relationality, rather than the ostensible skills of the social engineer, to better understand how these deceptions work.

While the social engineer works hard to develop a recognizable pretext, there comes a moment when the pretext has to be put into action. This is the moment of engagement with the wily target. This is a moment for bullshit. We turn to this next.

5
Bullshitting: Deception, Friendliness, and Accuracy

> This character is often so friendly, glib, and obliging that you're grateful for having encountered him.
> —Kevin Mitnick on the social engineer[1]

> ... Any editor will be assisted most cheerfully in verifying directly any statement of fact.
> —Ivy Lee, "Declaration of Principles"[2]

Social engineers are master bullshitters.

Some of them would even admit this openly. The phone phreaks of the 1970s through 1990s were especially open about being bullshitters. In fact, instead of calling it "social engineering," more often they simply called their practice of conning telco employees out of information "bullshitting."

Much like "trashing," the phone phreak term "bullshitting" sounds crude. No wonder contemporary security researchers use the phrase "social engineering"—it just sounds more respectable. But like trashing, bullshitting is a rich and complex concept. Focusing on it illuminates more than just telephone exploration. Using the concept of bullshitting as a lens reveals a great deal about contemporary interpersonal social engineering practice, and as well as the older mass social engineering practices of the early to mid-twentieth century. And looking ahead, this concept will join trashing and pretexting as tools to help us illuminate the contemporary practice of masspersonal social engineering.

Bullshitting is the central act of social engineering. It's what happens when the social engineer engages with the target. Social engineers need to trash their targets and learn everything they can about them. They then can construct pretexts—for example, ready-to-hand social stereotypes—that they can fully embody and that their targets will recognize. But all of that preparation will be for naught if the social engineer can't bullshit. There comes a moment when the social engineer must put their plans and technical knowledge to the test during the engagement with the target. This is a difficult moment, a trial of the social engineer's ability to be adaptive. During the engagement, social engineers rely on a peculiar, truth-indifferent mix of deception, friendliness, and accuracy to successfully manipulate their targets. That is what we mean by bullshit.

To understand bullshitting, we have to return to the phone phreaks who developed bullshitting as a social engineering technique. We'll join the phreaks in taking bullshit seriously by drawing on a small area of philosophical and sociological inquiry—the meaning and uses of bullshitting—to clarify what social engineers do when they meet their target. We will then take a look at phone phreak bullshitting guides and come to recognize bullshit for what it is: a skillful, truth-indifferent mix of deception, accuracy, and friendliness. We'll once again take this phreak and hacker concept and look back at the older, mass social engineers and find that they, too, were consummate bullshitters.

On Bullshit with the Phone Phreaks

We owe a great deal to the phone phreaks, who bequeathed us the theoretically rich concept of bullshit. Recall that the phone phreaks were people obsessed with the Bell telephone system in the 1960s through the 1980s. In order to explore its furthest reaches, the phreaks found ways to make free long-distance calls. The histories of and documentaries about the phone phreaks tended to emphasize the technologies they used to explore the Bell System, especially their "blue boxes" that could create special control tones that could manipulate the network. They are often lauded for their technical abilities to connect phone circuits together (a process called *tandem stacking*) and their skills in getting free long-distance calls. Less discussed are the phreaks' social skills in bullshitting the operator.

In its simplest form, bullshitting involves talking to telephone company employees, typically operators and technicians, in order to manipulate them into doing something for the phreak or giving the phreak access to vital internal information. This is the central activity of social engineering. A phreak might bullshit an operator, for example, and convince the operator to connect the phreak to a special phone line so that the phreak can make free long-distance calls. Another phreak might bullshit a phone technician in order to learn how a new switching system works.

To be effective, phone phreak bullshitting required a lot of accurate, technical knowledge—the sort of thing the phreaks would find during trashing runs in the dumpsters outside telephone company offices. But, of course, bullshitting was deceptive: the phreaks would often use pretexts, such as being a fellow Bell employee. And yet, the phreaks weren't cruel when they sought to deceive operators; in fact, they were downright friendly. Bullshitting then was far from simple: it was a complex mix of deception, kindness, and technical mastery, all brought together in calculated, interpersonal manipulation.

Despite the term sounding crude, the phreaks were onto something when they called their engagements bullshit. After all, our society is full of bullshitters, and this has attracted the attention of serious academic thinkers. What emerges from this scholarly analysis are two seemingly distinct conceptions of bullshitting. On the one hand, bullshitting is a harsh, dangerous, manipulative practice that undermines trust in institutions and deceives the audience. On the other, bullshitting is a social practice that allows for identity-play, social experimentation, and camaraderie. As we will see, both of these perspectives will hold when we consider phone phreak bullshitting as well as social engineering in all its forms. In the hands of social engineers, both of these conceptions merge in a creative, truth-indifferent mix of deception, accurate knowledge, and sociability.

Bullshitting as Indifference to Truth
Since the mid-1980s, there has been a small but robust group of philosophers developing a theory of bullshitting as it relates to lying, truth-telling, and deception. The debates are lively and address many important questions, including how to identify bullshitting, distinguishing bullshitting from lying, and considering the extent of bullshitting in contemporary society. The dean of this school of philosophers of bullshit is Harry Frankfurt, whose popular 1986 essay (republished in 2005 as a bestselling book), *On Bullshit*, kicked off the debate about the definition, practices, and extent of bullshitting in contemporary American culture.

Perhaps the most quoted line from *On Bullshit* is Frankfurt's point about the "essence of bullshit": bullshit is essentially a "lack of connection to a concern with truth . . . [an] indifference to how things really are."[3] For Frankfurt, bullshitting is far more dangerous than lying, because at least the liar is concerned with the truth or how things really are—even if they want to tell us the opposite of the truth. Instead, bullshitting is indifference towards truth or falsity.

"It is impossible for someone to lie unless he thinks he knows the truth," Frankfurt argues. "Producing bullshit requires no such conviction."[4] Thus, the bullshitter

> is neither on the side of the true nor on the side of the false. His [sic] eye is not on the facts at all, as the eyes of the honest man and of the liar are, except insofar as they may be pertinent to his interest in getting away with what he says. He does not care whether the things he says describe reality correctly. He just picks them out, or makes them up, to suit his purpose.[5]

Ultimately, for Frankfurt, the bullshitter's purpose is to deceive us as to their true intentions. One of his key examples is readily understandable: the politician who blusters on and on about our Glorious Country and its Special Blessings from God. This politician, Frankfurt argues, is far less concerned about true and false and far more concerned about our impressions of them as a God-fearing patriot. In this sense, this aspect of bullshit resonates with our discussion of recognition of pretexts in the previous chapter. Bullshitters can rely upon stereotypes and other easily recognized social practices as a deceptive cover.

Frankfurt's initial definition of bullshitting as an indifference to truth or falsity has been challenged by several other philosophers who develop counterexamples and offer technical arguments about situations where one might care about truth but still be bullshitting.[6] The extent of this debate is beyond the scope of this chapter, but for our purpose of considering bullshitting among social engineers, one modification of Frankfurt's theory is quite useful. It comes from the philosophers Andreas Stokke and Don Fallis, who modify Frankfurt's indifference to truth or falsity thesis into indifference towards the *inquiry* into what is true or what is false.[7] As they explain,

> We characterize bullshitting as a mode of speech marked by indifference toward contributing true or false answers to [questions under discussion]. The kind of indifference toward truth or

> falsity that characterizes the phenomenon of bullshitting is not indifference toward the truth-value of what one says, but indifference toward the effect that one's contributions have on the discourse.[8]

For Stokke and Fallis, then, bullshitting undermines the inquiry towards truth as this inquiry plays out in specific conversational situations they call "questions under discussion." These questions under discussion might be things like: when does the bus arrive? What's the weather like outside? Do you have much homework to do? Whereas a liar would know the truth and tell its opposite, and the honest person would simply answer truthfully to the best of their ability (including saying they don't know the answer), the bullshitter simply makes statements with no intention of moving the discussion toward truth. Instead, the bullshitter is more interested in impression management: perhaps evading the question, hiding true intentions, trying to appear friendly while really being uninterested, or the like. Thus, for Stokke and Fallis, "the bullshitter makes contributions while not caring about their effect on particular subinquiries."[9]

Setting aside the debates about definitions, if there is anything to fault in Frankfurt, Stokke, and Fallis's conceptualizations of bullshitting, it is that they underplay what is at stake when bullshitters bullshit. Who cares if the politician bloviates or the person bullshits about the weather with no evidence? But, as criminologist Daniel Mears notes, bullshitting can be part of exploitation and manipulation: "Bullshitting can provide a means by which to influence or control perceptions of reality and in turn with a means to achieve specific social, political, and economic goals."[10] This point is especially important when we consider social engineering.

Shooting the Bull
But before we turn to the instrumental, manipulative communication goals of social engineers, we ought to consider the more

sociable connotations of bullshitting, found in phrases like "shooting the bull" or "bullshit session."

Perhaps the most extensive consideration of this connotation of bullshitting is in communication and sociology scholar Chandra Mukerji's late 1970s work on hitchhikers. Much as Frankfurt would note almost a decade later, Mukerji argues that an indifference to truth is a key feature of bullshitting. But rather than see it as purely dangerous, Mukerji sees it as a sociable practice: "truth and falsehood are not issues in bullshitting because this kind of talk is playful; it is a way to make conversation more fun."[11] Hitchhikers, she argues, engage in playful bullshitting in order to reconstruct their experiences and hence themselves as heroes in "road stories." These stories present the hitchhiker as "worldly" with "enough guts, daring, endurance and friends to 'make it' on the road."[12] Moreover, the hitchhiker's audience—typically fellow hitchhikers or the people giving them rides—have a stake in such stories. Their stories are told to others who "have vested interests in glamourizing life on the road."[13] Given that these are stories vocally told to strangers, the stories could be forgotten or denied later, reducing the truth-value stakes while placing more emphasis on the sociable, entertaining qualities of the bullshit.

The criminology scholar Mears develops this sociable, identity-play aspect of bullshitting further, noting that key functions of bullshitting include socializing, exploring the self, expressing feelings, and passing time. As Mears notes, "To those well-versed in the art of bullshitting, there is an ability to define oneself and to achieve particular goals."[14] Thus, beyond hitchhiking road stories, we can imagine many social settings in which truth or falsehood are set aside in favor of plausible yet edgy conversation, from fishing stories to barroom tales to hallway talk at academic conferences.

To draw these two streams together—the indifference to inquiry towards truth and the sociability of "shooting the bull"—we can return to Frankfurt. Specifically, Frankfurt notes the artistry and creativity of bullshitting, connoted in the phrase "bullshit artist." As he explains,

> A person who undertakes to bullshit his way through has much more freedom [than the liar]. His [sic] focus is panoramic rather than particular. He does not limit himself to inserting a certain falsehood at a specific point, and thus he is not constrained by the truths surrounding that point or intersecting it. He is prepared, so far as required, to fake the context as well. This freedom from the constraints to which the liar must submit does not necessarily mean, of course, that his task is easier than the task of the liar. But the mode of creativity upon which it relies is less analytical and less deliberative than that which is mobilized in lying. It is more expansive and independent, with more spacious opportunities for improvisation, color, and imaginative play. This is less a matter of craft than of art. Hence the familiar notion of the "bullshit artist."[15]

For Frankfurt, bullshitting is playful and dangerous, sociable and subversive, creative and cunning. Such an artistic approach can serve the purposes Mukerji identified: to have fun, create entertaining stories, build relationships, and foster community. It can also be used to undermine a drive towards truth, as Mears, Stokke, and Fallis observed. Truth, indeed, is beside the point.

Exquisitely Accurate Bullshit

And yet, we must emphasize that bullshitting is not lying. It is not the simple telling of the opposite of the truth. In fact, bullshitting often involves getting details, facts, and arguments right—at least to the extent that they serve the bullshitter. Frankfurt points to entire industries, such as advertising and politics, and notes the

> exquisitely sophisticated craftsmen who—with the help of advanced and demanding techniques of market research, of public opinion polling, of psychological testing, and so forth—dedicate themselves tirelessly to getting every [bullshit] word and image they produce exactly right.[16]

Indeed, the idea of *caring* for the bullshit—for getting it just right—reflects an older meaning of "accuracy": from the Latin *accuratia*, or "care, attention." The underlying research and planning that bullshitting often requires is tremendous: bullshitting works best when the target is extensively studied, the language of the target is fully understood, and the messaging is carefully calibrated to reach that target. (Otherwise, why would any phreak or hacker jump into dumpsters?)

Like Frankfurt, Mukerji observed hitchhikers caring for the bullshit they provided. Their stories of the dangers of the road must be "plausible . . . too much exaggeration can make the story implausible."[17] Plausibility relies upon articulating the bullshit story with the audience's experiences. An accurate story is one that resonates with the audience's own understanding of life on the road; a misalignment means that storyteller's bullshit is called out.

Thus, facts can matter—even in a bullshit story. The bullshitter's contributions can be true statements, accurate information, or deception—the point is that this *mix* of lies and truth are made with indifference to the understood purposes of the conversation. In the case of bullshitting the operator, the phone phreaks undermine the operator's "questions under discussion" not with pure lies but with a careful mix of accurate information and deception that covers up the phreaks' intentions.

Bullshitting the Operator: Best Practices

Here, we want to return to our benefactors, the phone phreaks, who generously gave us the rich concept of bullshit.

The three-fold conceptualization of bullshitting as an indifference towards the inquiry into truth, a playful, sociable practice, and an act of care and attention to information, maps back onto the phone phreak guides to bullshitting the operator. These guides can be found in a range of underground magazine articles (such as *2600* and *TAP*), computer Bulletin Board System (BBS) posts, and (to a lesser extent) mainstream press articles appearing in the late 1970s to mid-1980s. These guides reveal the exquisitely sophisticated craft of phone phreak bullshitting as a mix of deception, sociability, and getting it right. They demonstrate how bullshitting can amplify the social engineer's deceptive *pretexts*, why *being friendly* is the social engineer's best bet, and how to *use the lingo* gathered during trashing. These practices are precursors to contemporary, masspersonal social engineering practices, and they also throw new light on the older mass social engineering.

Deception: Pretexting

Phone phreak guides to bullshitting often recommend that the would-be bullshitter come up with a plausible role to play before calling up the Bell operator. This is of course the practice that would eventually become formalized and referred to as "pretexting," although the phreaks themselves did not use that term. Pretexts give the bullshitting phreak an advantage as they seek information or access, because they set the stage for the subsequent conversation.

The guides get quite specific about titles and ways to introduce oneself to the operator. An early guide in *TAP* recommends pretexts such as being a fellow operator or Bell security employee.[18] A BBS guide with the enigmatic title "Flying Penguin Presents: Bullshitting the Operator" provides even more roles to play. When bullshitting an operator, Flying Penguin advises us, "we must be from the phone company."[19] To that end, the guide offers a list of titles the phreak could adopt, including

- Toll service maintenance engineer
- Station repair
- Cable MTCE technician
- TSPS maintenance/maintenance administrator

- Central Office Supervisor
- TSPS Security[, and]
- Toll Service Maintenance.[20]

Beyond roles, the guides offer scripts. A guide in *TAP* provides a script for talking to Customer Name and Address (CN/A) operators:

> Hi, this is Jim with the Residence Service Center in Pasadena. I've got a customer on hold who claims not to have made a rather lengthy call to (212) 555-1212, so I need the customer's name on that please. That was (212) 555-1212.[21]

Likewise, "Sharp Remob's Guide to Bullshitting the Phone Company Out of Important Information" suggests calling the operator and saying, "Hi, this is Bob Dwyer with repair, do you show any order activity on 555-2344?"[22]

Pretexing is the most clearly deceptive aspect of phone phreak bullshitting. To return to the philosophers Stokke and Fallis, a phreak claiming to be someone they are not completely undermines the specific inquiry into truth undertaken by the operator. Arguably, the use of a pretext is a lie, maybe even an Augustinian lie: the phreak knows perfectly well they are not who they claim to be. The operator answers the phone with the intention of helping a fellow Bell employee or customer, and the phone phreak is only too happy to maintain this illusion during the inquiry in order to gain access to information or other resources. But even the best pretext cannot function without other elements of bullshitting—including offering a bit of friendliness.

Being Friendly

Bullshitting an otherwise naive operator or clueless technician in order to manipulate them or gain even more information sounds pretty harsh. This is part and parcel of the connotation of bullshitting as deceptive indifference to truth or falsity. However, returning to the jovial side of bullshitting (as we saw in Mukerji's study of hitchhikers), the phone phreaks did not just manipulate the operators; they were friendly to them.

An article in *TAP* reminds the reader that "operators are people too, y'know. So always be polite, make good use of 'em, and dial with care."[23] Likewise, the Flying Penguin's "bullshitting the operator" guide reminds phreaks to "always say thank you!"[24] Another phreak recommends sounding "friendly and natural."[25]

An illustration of how friendliness functions can be seen in one of the rare recordings of an actual social engineering engagement available. In front of an audience at the first HOPE hacker conference in 1994, a hacker going by the handle SN calls Sprint, looking to get a Customer Name and Address (CN/A) telephone number—a phone number that only Sprint employees should have access to. He bullshits a Sprint operator, Deborah Brown (DB).

DB: Sprint Customer Service, this is Deborah Brown speaking, and how may I help you?

SN: Hey, Deborah, how's it going?

DB: It's fine.

SN: This is Bob Dwyer, over at Spring Social Engineering, how you doing?

DB: I'm fine. And yourself?

SN: Pretty good, you know, it's one of those days.

DB: Okay.

SN: Umm, you wouldn't happen to have the number for the CN/A that handles 313? That's Michigan.

DB: I think so . . . just a moment. (humming, computer keyboard tapping)

SN: So, you guys busy over there today?

DB: We were *real* busy earlier.

SN: Yeah, I know *exactly* what you mean.

DB: Ok, it's 313 . . .

SN: Mm huh.

DB: 424 . . .

SN: Uh huh.

DB: 0900.

SN: OK. . . . Alright. Hey! Is CIS up for you guys over there?

DB: Yes.

SN: Yeah? 'Cause we've been having a lot of problems with data loss from CIS to the switch, and it's not processing the TCs too well. What do you show is the last account that you, umm, handled or processed?

DB: Last account?

SN: Yeah.

DB: Umm . . . 18–

SN: Oh, wait, we have it right here. Yeah, in fact, CIS just came back online for us.

DB: Did it?

SN: Yeah.

DB: Oh.

SN: Hey! Thanks a lot for your help!

DB: You have a good one!

SN: Alright, bye-bye!

DB: Goodbye![26]

SN is very careful to include sociable lines: "how's it going?", "you guys busy over there?", and "you have a good one!" He isn't just engaged in an instrumental hunt for information; his friendly demeanor and kindness to his supposedly fellow employee helps smooth the interaction (and, perhaps, distracts from his audacious self-identification as "Bob Dwyer from Sprint Social Engineering").

Such friendliness is a great tactic when bullshitting targets at work. As a BBS post titled "The Official Phreaker's Manual" explains,

> Most Bell employees are really glad to talk to someone. Remember, they usually interact with disgruntled customers with complaints. Their spouses probably yell at them, and their supervisors either complain about their performance or ignore them. Society at large just doesn't care about them. They're most probably disenchanted with the world at large, and maybe even dissatisfied with their jobs. The chance to talk to some one who merely wants to listen to what they say is a welcome change. They will talk on and on about almost anything, from telecommunications to their home life and their childhood. The possibilities for social engineering are endless. Remember, Bell employees are humans, too. All you have to do is listen.[27]

This is bullshitting in the social, "shooting the bull" sense that Mukerji observed among the hitchhikers. Friendliness, a bit of camaraderie in a cold, corporate setting makes the interactions smoother and the information flow more freely.

But the admonition to be friendly may not be as much about humanity and compassion as it is about improving the phreaks' control over the Bell System. A *TAP* article on bullshitting CN/A operators notes the utility of friendliness: "When the employee [read: phone phreak] sounds natural and cheery, the CN/A operator doesn't ask any questions."[28] Here, being friendly is thus more than mere politeness; it helps the bullshitting operation by covering up the phreak's intentions to gain illicit access to information behind a cloud of joviality and kindness. This reflects something that communication researchers who study deception have found: friendliness is associated with "honest" demeanor, whether or not the friendly person is telling the truth.[29] Friendly bullshit keeps intrusive questions at bay.

Moreover, friendliness has the side effect of helping preserve bullshitting possibilities for future phone phreak use. As one phone phreak advises,

You SHOULDN'T, if you screw up, or if the phone co. employees are uncooperative, break down and swear at them or call them names. This will only contribute to the destruction of these departments for engineering purposes.[30]

Likewise, a rare guide to *in-person* phone phreak bullshitting describes how to get a tour of a Bell Central Office and urges anyone who engages in such a tour to "Make sure to leave a good impression so that fellow telecommunications hobbyists can tour the place in the future."[31]

Accuracy: Using the Lingo

However, a friendly tone will only take you so far. This is especially the case when the phreak is pretexting as a fellow telco employee. Such a role is only possible if a phreak knows the Bell System language. The bullshitting guides of the phreaks heavily stress learning the Bell system lingo and using it accurately.

For example, a 1974 guide to bullshitting a special class of phone operators explains that knowing the lingo lends an "air of authenticity" when dealing with these operators.[32] Here, being "authentic" means using the right terms, such as these suggested lines: "This is Phil Donehue on the 4-A 17-C test board. We have some trunk testing to perform and require a no-test trunk for the 555 office."[33] With the right lingo spoken to the right operator, this gives the phreak a, well, freaky ability: the ability *to listen to other people's phone conversations*. "The fact is that just about anyone with the right numbers to call and the correct things to say could tap into anyone's telephone line using telephone company circuits."[34]

Another guide also makes a similar argument in an article on "Rate & Route" operators, a special class of telephone operators. "The [Rate & Route] operator has a myriad of information, and all it takes to get this data is mumbling cryptic phrases at her."[35] The guide goes on to share the "cryptic phrases": "numbers route," "directory route," "operator route," and "place name," to name a few. The most extensive phone phreak guide is probably BIOC Agent 003's "Course in Basic Telecommunications," a series of text files appearing on BBSs around the years 1983 and 1984.[36] These files include information on bullshitting the operator as part of a larger, exhaustive emphasis on the technical terms and details of the phone system itself. The guide explains how the fruits of trashing and research can lead to better social engineering engagements.

As these guides show, the phreaks go to great lengths to learn the terms and use them correctly. Thus, in contrast to the deception of the pretext, the phone phreak's use of lingo demonstrates the careful use of accurate information during the bullshitting operation. Like Frankfurt's "exquisitely sophisticated craftsmen" or Mukerji's hitchhiker who spins a plausible tale, the phreaks work to get things right.

But again, these deployments of correct information are not in service of the operator's understanding of the goals of inquiry, but instead are used to cover up the true intentions of the phreak. If we speak of these practices purely in terms of "true" and "false"—if we use terms like "lies" or "fake news"—we might say that the bullshitter is concerned with truth only so far as it relates to the effectiveness of the communication. Whether the bullshitter is aware of the truth or not, they will likely attempt to achieve verisimilitude sufficient to achieve the desired effect. The bullshitter might not know or care about the truth, but will likely want the audience to at least see the bullshit as plausible. This is what we mean by accuracy. Despite the accuracy of the phreaks' use of phone lingo, the care and attention paid to getting things just right, they are still bullshitting.

Bullshitting in Hacker Social Engineering

The contemporary, professionalized field of interpersonal hacker social engineering no longer uses the term bullshitting. Indeed, the transition from the less polite "bullshitting" to the more professional-sounding "social engineering" occurred in the 1980s through 2010s. Sometimes, the terms "bullshitting" and "social engineering" appeared alongside one another in BBS posts or magazine articles. Eventually, however, "social engineering" replaced "bullshitting" as the term of art among hackers.

Something gets lost when "bullshitting" is dropped from the lexicon. Bullshitting is an excellent label for what happens when social engineers leave the trashing/OSINT stage to put their pretexts to the test in the field and engage with people. Hacker social engineers may not call it bullshitting, but they are consummate bullshit artists, nonetheless. As the hacker social engineer Kevin Mitnick puts it, "successful social engineers have strong people skills. They're charming, polite, and easy to like—social traits needed for establishing rapid rapport and trust."[37] Contemporary social engineering guides, such as the books by Hadnagy, Conheady, or Mitnick, build on the phone phreaks' older guides by including discussions of role-playing and deception (in the form of developing a pretext), accuracy (in the form of information gathering and getting terms, names, and jargon right), and friendliness. On this last point, Johnny Long's *No Tech Hacking* starts with an anecdote about his first physical social engineering engagement, when his mentor suggested Long enter the building via a loading dock. When Long protested—"there's people there!"—his mentor prescribed bullshitting: "Just look like you belong. Say hello to the employees. Be friendly. Comment on the weather."[38] Indeed, an unofficial motto that professional social engineer Chris Hadnagy offers his students is: "leave them feeling better for having met you."[39]

This mode of bullshitting, adopted from the phreaks, is predominantly aimed at interpersonal manipulation. Like the phreaks before them, contemporary social engineers might use "vishing" (a portmanteau of "voice" and "phishing") to ask for passwords over the phone. Or they may use bullshit in emails as they phish. And they also bullshit in person when they go to physically penetrate a target organization.

Bullshit among the Mass Social Engineers

Bullshit isn't limited to the interpersonal social engineering of phreaks and hackers.

If there's any group of social engineers who never used the word "bullshit" (at least in print), it's definitely the mass social engineers of the early to mid-twentieth century. Indeed, their adaptation of the Progressive-era emphasis on "facts" means they would be incensed to have their techniques of controlling crowds through media messages called "bullshit."[40] Phone phreaks and hackers may have been comfortable with such crude language, but not mass social engineers like Bernays, Fleischman, or Ivy Lee.

But as we have shown, "bullshit" is a rich theoretical concept. Much like the phreak and hacker term "trashing," which also was never used by mass social engineers, it's profitable to take up bullshitting as a lens to reconsider the activities of the mass social engineers. Indeed, arguably Harry Frankfurt had their ilk in mind when he composed his book *On Bullshit*. His sharpest condemnations are directed at the use of bullshit by marketers and political communicators.

And those condemnations seem justified. From the start of his career, mass social engineer Ivy Lee peddled bullshit. In 1914, he was hired by the industrialist John D. Rockefeller, Jr., to advocate on behalf of the family after a mining company the Rockefellers had invested in paid an armed militia to break a union strike earlier that year. The militia slaughtered union members in the town of Ludlow, Colorado, including women and children who tried to hide from the militia under a tent, but were killed when the tent was set on fire by the strike-breakers.[41] The event is now known as the Ludlow Massacre.

The Rockefeller family didn't need more bad publicity; thanks to the work of the muckraker Ida Tarbell, they were already vilified due to their history as the owners of the legendarily rapacious Standard Oil corporation.[42] Their involvement in a bloody, machine gun-driven, tent-burning, union-busting operation was not helping their reputation. To help clear the Rockefeller name, Lee sent tens of thousands of bulletins and booklets "to opinion leaders throughout the country: 'public officials, editors, ministers, teachers, and prominent professionals and business men.'"[43] Conceived of as missives exploring the "struggle for industrial freedom," these bulletins included benign facts about Colorado coal mining meant to inform people about the industry. They also included material meant to sully the unions and strikers. As the media studies scholar Stuart Ewen notes, Lee mixed accurate facts about the industry with "calculated inaccuracies—Lee's dispatches, for example, routinely exaggerated the salaries received by the union organizers."[44] A biography of Lee concurs with Ewen's assessment, explaining that "most of the bulletins contained matter which on the surface was true but which presented the facts in such a way as to give a total picture that was false."[45] This mix of fact and deception was designed to prove the bullshit claim "that the pillage at Ludlow was the work not of the mine operators and their armies, but of 'well-paid agitators sent out by the union.'"[46]

Moreover, Lee used deceptive pretexts. The bulletins were labeled as being produced by the Colorado coal industry, with no mention of the Rockefeller family's sponsorship.[47] He also recruited one Helen Grenfell, the "Vice-President of the Women's Law and Order League of Colorado," to provide an "eyewitness" account of the Ludlow Massacre that blamed the union for starting the fight and stated that the fatal fire was caused by accident, not set deliberately by the strike-breakers. "Unmentioned in the report were the facts that Grenfell was not, in fact, an eyewitness to events at Ludlow and that she was the wife of a railroad official whose company profited from carrying Colorado coal."[48]

The use of deception alongside facts contradicted Lee's stated ethical standards. Writing in his 1925 pamphlet *Publicity*, he declared that "the essential evil of propaganda is the failure to disclose the source of information."[49] In his "Declaration of Principles," a now-famous statement he had sent to newspaper editors across the United States just nine years prior to the Ludlow Massacre, Lee assured wary news editors that his public relations work "is accurate," and that "any editor will be assisted most cheerfully in verifying any statement of fact. . . . Full information will be given to any editor concerning those on whose behalf an article is sent out." His goal is to "supply to the press and public of the United States prompt and accurate information concerning subjects which it is of value and interest to the public to know about." Lee concluded with a promise: "I am always at your service."[50]

Apologists for Lee's work on behalf of the Rockefellers—with its pretexts and bullshit about union leader salaries as well as who started the conflict—note that Lee could not possibly have lived up to his own declaration, because he was fed information from his employer and from a journalist in Colorado and he didn't have the time to verify every detail.[51] However, when Lee was called out on the deceptions in his work by the US Commission on Industrial Relations, he did little to correct the record, despite his promise of cheerful assistance. To Lee's credit, he did send out some corrections in the form of one thousand booklets. But this was a tiny fraction of the tens of thousands he had sent with the deceptive information.[52] Moreover, "despite Lee's efforts to rectify the problem, copies were still being circulated without the correction."[53] And none of this addresses the fact that Lee did not disclose that the Rockefellers were behind his bulletins, belying his promise to provide "full information . . . concerning those on whose behalf an article is sent out." And although Lee was chastened when his bullshit was called out by the Commission, he repeated his tactics on behalf of coal mine operators in West Virginia just a few years after Ludlow.[54]

Lee's work was quintessential mass social engineering bullshit—glorying in facts and accuracy when presenting statistics about Colorado coal mining, and yet deceiving readers about his sources and peddling inaccuracies that supported his public relations mission. The truth of the Ludlow Massacre was beside the point. The point was to sway public opinion towards the Rockefellers and away from the union, and to do so behind the ethical veneer of a declaration of principles.[55] But Lee did adhere to at least one part of his principles: he was, by all accounts, cheerful and friendly. John D. Rockefeller Jr. commended him for doing his work in an "entirely good natured, attractive, and impressive manner."[56]

Despite Lee's bullshitting, or perhaps because of it, he is now considered a foundational figure in the field of public relations. To this day, public relations professionals point to his "Declaration of Principles" as the start of the modern field.[57] However, American studies scholar Jonathan Auerbach notes that, while Ivy Lee has the best claim to the title "father of public relations," most scholarship gives that honor to Edward Bernays. This is because Bernays

> was happy to put himself forward as the father of public relations, even though he clearly came on the scene a full decade after Lee. Considering influence rather than seniority, I would again insist that Lee remains the more important figure in the history of public relations, despite Bernays's relentless efforts to toot his own horn. The simple fact is that Bernays was more fun and flamboyant than Lee. He was also a far better self-promoter and propagandist for propaganda.[58]

In other words, Bernays was an even better bullshitter than Lee, a charming and affable character who bullshitted himself into the limelight as the "father of PR."

Multiple scholars of Bernays note that he exaggerated the success of his own campaigns. For example, over the course of his long life, he overstated the impact of his Torches of Freedom march, the event he and Fleischman concocted for the American Tobacco Company.

The Torches march included a few women smoking while marching in the 1929 Easter Parade in New York City in order to, in his words, "smash the taboo against women smoking."[59] In Bernays's telling of his life, the event went from being a minor one to the most transformative moment in American tobacco history, being covered in every major newspaper and being the single reason why women smoked in America.[60] In a video interview in the Museum of Public Relations, a grandfatherly Bernays claims that the day after the Torches of Freedom march, "there wasn't a newspaper in the United States [ignoring the story.] Even the *New York Times* had a front page story, 'Debutantes Light Torches of Freedom To Protest Man's Inhumanity to Women by a Taboo Against Smoking.'" While it's accurate to say that the *Times* had a front-page story on the Easter parade, the headline was, in fact, "Easter Sun Finds The Past In Shadow at Modern Parade." The phrase he quotes is non-existent, and the smoking women are mentioned only in passing.[61] In addition, Bernays was also wrong about the impact of the event: women had smoked in public for years prior to the march and the coverage of the event was far less extensive than Bernays claimed.[62] Bernays obviously had a stake in our believing in his mass social engineering prowess, so he bullshitted about it.

Most strikingly—despite his pseudo-feminist campaigns to help women enjoy the freedom of smoking—Bernays's drive to be the "father of PR" led to him completely erasing the work of someone who might claim the title "the mother of public relations." His wife, Doris Fleischman, gets even less credit than Ivy Lee for her role in developing the field, despite the fact that she was an equal partner in their public relations firm and did much of the key work in developing theories of mass social engineering.[63] Bernays's reason for not giving his partner her deserved credit? There was a taboo on women in professions, and his clients wouldn't accept Fleischman.[64] So much for taboo-smashing.

The rampant bullshitting among mass social engineers may be why Frankfurt was unsparing in his criticism of them:

> The realms of advertising and of public relations, and the nowadays closely related realm of politics, are replete with instances of bullshit so unmitigated that they can serve among the most indisputable and classic paradigms of the concept.[65]

While the phreaks were at least honest in calling what they did "bullshit," the mass social engineers are the ones most associated with the practice in modern American thought, mixing deception, facts, and friendliness together on a mass media scale. Lee and Bernays are, to this day, celebrated figures in the field of public relations. We owe them a debt of gratitude for their finely crafted bullshit.

Conclusion

Social engineers—mass, interpersonal, or otherwise—are bullshitters. Social engineering is an artistic, truth-indifferent mix of deception, accurate information, insider knowledge, friendliness, talking out of one's ass, flattery, and self-aggrandizement. Bullshitting offers a flexible suite of communication skills for the social engineer. It is sociable, but with instrumental goals: to keep the pretext alive and to get valuable information, access to restricted technologies or networks, or control of crowds through media messaging. Bullshitting, in sum, is the social engineer's engagement tactic of choice. Whether it is a phone phreak calling up an operator, a hacker social engineer sending a phishing email, or a public relations firm creating a media campaign, bullshit is always involved. Bullshit is dangerous, capable of radically undermining our trust.

So, if social engineering relies heavily on bullshitting, why don't we throw the social engineers out? The answer is simple: they claim to be doing it for our own good. We turn to that next.

6
Penetrating: The Desire to Control Media and Minds

> Remember—once you are a social engineer, you deceive, manipulate, and trick people for a living . . . but you also educate them.
>
> —Sharon Conheady[1]

"I just wanna F you up," says hacker social engineer Jayson Street at DEF CON 19. "I just wanna mess you up in the worst possible way. I wanna be the worst thing to ever happen to you at the worst possible time."[2]

Street is presenting his approach to hacker social engineering, specifically what professional hacker social engineers call "penetration testing" or "pentesting" for short. He has stolen purses, phones, documents, laptops, even cars, all while being paid to do so as a professional pentester hired to seek flaws in an organization's security. His talk, titled "Steal Everything, Kill Everyone, Cause Total Financial Ruin" sounds vicious and prurient. But it's in service to a greater goal: Street is hired by corporations to test their security. He's not an underground hacker—he's a professional. Part of the process involves educating the corporations he's "F'ed up" on their vulnerabilities. After conducting his tests, he produces reports meant to teach lessons about security to the hapless corporate employees who assume they're safe from interpersonal social engineers.

Professional penetration testing of organizations and their information systems has been formally practiced since at least the mid-1960s.[3] It can be done over networks and via software, and much of the early literature on pentesting focuses on seeking out software exploits in computer operating systems.[4] Our interest here is, of course, the social engineering iteration of pentesting—the manipulation of people in order to gain access to a system to "F it up." An early example of hacker social engineering in professional pentesting occurred in 1985, when NASA hired a computer security firm to test the security of the Goddard Space Flight Center. The security firm used a combination of software and social engineering attacks. Like so many others, the security team found that social engineering was an extremely effective method for penetrating security systems.[5]

An entire industry of social engineering penetration testing services is now available for hire. Security consultants like Jayson Street, Sharon Conheady, Jenny Radcliffe, Johnny Long, Chris Hadnagy, and Kevin Mitnick specialize in using social engineering to break into corporate systems and buildings. Corporations that don't want to hire outside consultants may opt to hire their own internal "Red Teams" to run regular pentests.

Like the other hacker social engineering terms that we're exploring in this book, *penetration* has a rich set of connotations, making it a powerful lens through which to look at other forms of social engineering. We'll start, as usual, with an overview of how interpersonal hacker social engineers have used this term, and then turn our attention to how mass social engineers also rely on logics of penetration. We will see penetrating metaphors—specifically, male sexual conquest and bullets—as we trace both interpersonal and mass social engineering's desire for control through communication. We will consider how social engineering penetration was professionalized, complete with social scientific theories, standardized methodologies, and metrics for gauging success. And we will also see how all social engineers "F us up" not just out of a desire for domination, but also in order to educate the rest of us about the dangers of being penetrated. They penetrate us for our own good.

Interpersonal Penetration Metaphor: Sexual Conquest

The hacker term "penetration testing," writes Hadnagy in *Social Engineering: The Science of Human Hacking*, "opens itself up for a slew of non-humorous sexual innuendos." He goes on to decry pentesters who claim they have "raped" computer servers. "I do not find that statement funny," he says. Admirably, Hadnagy demands that his colleagues never use such language, noting that it will turn many people away from the field.[6]

But the fact is that hacker social engineers have long articulated "penetration" with sexual metaphors—even violent and misogynistic ones—and they will likely continue to do so. We cannot ignore this connotation. Once again, despite some of the crudity of hacker parlance, a term such as "penetration" reveals underlying logics. In this case, penetration as metaphorical sexual domination contains articulations of *technical mastery* and *control*, often presented in a hyper-masculinized manner. The hacker vision of social engineering as the interpersonal manipulation of other people heavily indulges in this articulation of technique, control, and masculinity.

As we have argued, hacker social engineering is the social side of hacking, belying the idea that hackers are antisocial loners sitting in darkened rooms lit only with the glow of computer screens. It also belies the idea that hackers rely exclusively on mastery of electronic computing technology. Instead, social engineering involves communicating with others. As such, hacker social engineering arguably involves abilities we might call "soft skills" or "people skills." Running the risk of gender essentialism, we may even say such skills are associated with feminized values.[7]

However, despite its possible associations with "soft" or even feminized social skills and its ostensible lack of *techne*, hacker social engineering is often presented as a highly rationalized and technical practice. As feminist scholars of science and technology have shown, values of rationality and technicity are often articulated with masculinity.[8] Such an articulation is very strong in hacking, because "hackers construct a more intensely masculine version of the already existing male bias in the computer sciences."[9] As for interpersonal hacker social engineering specifically, when we consider its underlying theoretical conceptions of sociality and human communication, we get a vision of other humans as knowable, transparent, manipulable objects, just as programmable as an electronic computer. Alongside this, the manipulator of the object—the hacker social engineer—is seen as a mindful, self-controlled, calculating subject. Say the right things at the right time, the hacker social engineer tells us, and you can get your target to do as you wish, a process US Naval Academy professor Joseph Hatfield aptly calls "technocratic dominance."[10] Social engineering thus sees humans as controllable objects, a means to the end of penetrating information systems. Masculinity, control, and social skill are articulated in hacker social engineering.

That this articulation appears under the label "penetration testing"—bringing with it the sexual innuendos Hadnagy decries—is not a historical accident. Interpersonal hacker social engineers have often associated with the hypermasculine world of pickup artistry. Pickup artists train themselves and other men on how to have sex with as many women as possible. As masculinity studies scholars argue, pickup artistry can be understood as "nerd masculine." Nerd masculinity values "rationality and technological proficiency," keeps women excluded, and draws on the logics of computer games, rule-bound spaces where the player qua avatar can achieve superhuman feats.[11]

As a nerd masculine field, pickup artistry shares affinities with computer hacking. More to the point, it has a shared history. A key example is Lewis De Payne. De Payne is notable for a variety of reasons: under his pseudonym Roscoe, he was the leader of the "Roscoe Gang," a Los Angeles-based group of phone phreaks and hackers that included Kevin Mitnick and Susan "Thunder" Headley, all of whom included social engineering in their toolkits and were written about extensively by journalists in the 1980s and 1990s.[12] But in addition to being a social engineer, De Payne is also notable for founding one of the internet's first pickup artist discussion forums, the Usenet newsgroup alt.seduction.fast, in the mid-1990s.[13] The alt.seduction.fast newsgroup distributed the teachings of Ross Jeffries, one of the founding fathers of the "seduction community" and a proponent of the 1970s-era psychological theory of neuro-linguistic programming (NLP).[14] The social engineer De Payne studied with Jeffries and became well versed in Jeffries' techniques of "speed seduction" in his own right in the 1990s.[15] Journalist Jonathan Littman verifies this in *The Fugitive Game*, even joining De Payne one afternoon to study Jeffries's seduction course. "It's the ultimate hack," Littman writes of De Payne's use of speed seduction, "talking women into going to bed with a computer nerd."[16] De Payne not only brought the teachings of Jeffries to the internet; he also authored his own book on seduction via computer Bulletin Board Services (BBSs), titled *Sensual Access: The High Tech Guide to Seducing Women Using Your Home Computer*.[17]

Later, in the early 2010s, the pickup artist/social engineer relationship was further strengthened by a cell phone phreak turned pickup artist Jordan Harbinger. In 2012, Harbinger joined Hadnagy's *Social-Engineer.org Podcast* to provide the perspective of a pickup artist in discussions of social engineering.[18] He was a member of the podcast for three and a half years. Prior to joining Hadnagy's podcast, Harbinger was the co-host of another podcast, *The Art of Charm*, which had been running since January 2007 and featured episodes such as "No More Mr. Nice Guy," "The Chemistry of Connection," and "What Women Think About Confident Men."[19] He was also the veteran of another podcast, *The Pickup*, and of a dating consultation talk show on Sirius Radio called *Game On*.[20] But like De Payne, Harbinger was not merely a pickup artist; he was also a phone phreak. His childhood hobby was exploring phone networks. He was fascinated by how cellphones worked and wanted to control them. He even posted information about cellphone hacking to the *2600* message board.[21] Thus, like De Payne before him, Harbinger brought knowledge from two domains, pickup artistry and hacking, to the *Social-Engineer.org Podcast*.

Penetration obviously takes on a literal meaning among the sex-obsessed men of pickup artistry. But the most relevant aspect of the pickup artist game is its emphasis on *controlling* others. Pickup artistry objectifies the other—in this case, women—and claims that the woman-object can be manipulated and controlled with the right behavioral stimuli. For a disturbing example, consider pickup artist Derek Rake's "Shogun Method" of "mind control."[22] Rake's goal is "emotionally enslaving" women, and he relies on a range of systematized verbal and nonverbal communicative techniques to do so. Pickup artistry relies heavily on the conceit that other people are programmable "neural machine[s]" who are thus vulnerable to control through interpersonal communication.[23]

Similarly, penetration among social engineers also emphasizes control of an objectified human. Hadnagy's books feature discussions of manipulating people's emotions—getting them to feel sympathy for him, or using fear—to get them to take the actions he wants.[24] In his first book, Hadnagy also endorses neuro-linguistic programming (NLP)—the psychological theory that pickup artist Ross Jeffries adhered to, even though NLP's vision of programmable humans has been repeatedly debunked.[25] Sharon Conheady's book *Social Engineering in IT Security* is a bit more sophisticated, drawing on social science to analyze authority, reciprocity, and mindlessness as a means to control the people she interacts with.[26]

Of course, a distinction between the all-male pickup artists and social engineers is that the latter field can include more than just men. Susan "Thunder" Headley is perhaps the most notable example, and she saw feminized sexuality as a powerful technique of control. In her DEF CON presentation in 1995, she recommends that women social engineers use the promise of sex to manipulate men.[27] More recently, with the identity-play possibilities of internet communication, social engineers of any gender identity can use this technique of control; a famous example is the Robin Sage experiment, where a fake "hot girl" persona was used to manipulate members of the defense industry.[28] The pickup artist and phone phreak Jordan Harbinger replicated the Robin Sage experiment on his own, using a fake LinkedIn profile based on his "gorgeous [female] assistant" to gather the personal information of people with top security clearances.[29]

Ultimately, success in penetration among interpersonal social engineers is the conquest of systems, such as computers or buildings. Hacker Johnny Long describes his joy when he is able to "have my way" with penetrated computers, downloading files, altering the contents of the server, or deleting it.[30] As for buildings, Jayson Street boasts, the "number 1 fact" is "I'm getting in, ok?" Social engineer Jenny Radcliffe reports she's gotten into "loads" of buildings, "too many to say."[31] Once inside, Street or Radcliffe can "F everything up" and cause the controlled chaos they are paid to create. Thus, the humans they control and manipulate are not the end goal: penetrating the system itself is. Here, too, penetration among interpersonal hacker social engineers echoes penetration among pickup artists: just as pickup artists are discouraged from settling down with just one woman—instead, their goal is to conquer as many as possible—hacker social engineers do not fixate on any given human.[32] All humans, for them, represent a means to an end: the penetration of the next system.

Professional Penetration

Professional pentesters don't control people and break into systems for free. They're hired to do so, often by corporations or organizations that are concerned about security. As much fun as it is to have their way with a system, it's still a job. Penetration testers work regular business hours to conduct their tests, write up security reports, and present their findings in meetings.[33] It's a far cry from the stereotypical, lulzy hacker underground of Mountain Dew-fueled 3 a.m. hacking, but it can be lucrative: the US Bureau of Labor Statistics reports that median pay for professional pentesters is around $100K.[34]

Pentesting's professionalization is reflected in changes in the hacker terms we've documented in this book. As we have shown, terms like "trashing" have been transformed into the professional-sounding "OSINT" (open-source intelligence). We've also noted that the more common name for hacker social engineering was once "bullshitting," a term now rarely used. Such transformations have been bolstered by formalized education and career titles such as "penetration tester." As of this writing, people interested in social engineering can take courses on the topic, including a master's level course at the University of Arizona, "MIS 566 Penetration Testing: Ethical Hacking and Social Engineering," or a variety of private instruction courses, such as Chris Hadnagy's "2-Day Social Engineering Bootcamp."[35] These courses are more than just learning how to bullshit: they encompass a whole range of theories, methods, and practices in order to produce professional social engineers who can be hired to conduct penetration tests. Expect to see more such courses, since the Bureau of Labor Statistics predicts a 32 percent growth in the information security sector over the next decade.[36]

The transformation of the crude terms of phreaks and hackers into terms acceptable in college classrooms and corporate boardrooms reflects the transformation of underground, vilified hackers into professional security consultants, enacting a "melodramatic arc" of the "idealized lifecycle of the hacker," where hackers reform, abstain from their previous illegal activities, and contribute to society by selling their skills in the marketplace.[37] Susan "Thunder" Headley is a pioneering example. As part of De Payne's "Roscoe's Gang," she regularly broke into Pacific Bell's systems in the late 1970s and early 1980s.[38] However, she transformed herself into a professional penetration tester, first appearing on ABC's *20/20* in 1982, instructing Geraldo Rivera on the finer points of hacking.[39] She then testified to the US Senate in 1983 and reportedly provided a social engineering penetration test to the US military.[40] After that, she offered her services as a professional pentester through the 1980s and 1990s before shifting careers to politics, poker, and coin collecting.[41]

Headley's transformation prefigured that of Kevin Mitnick, who would replicate Headley's trajectory almost exactly in the 1990s and 2000s. After serving his prison sentence in the 1990s, Mitnick also appeared on national television—in this case, CBS's *60 Minutes*—and also testified to the US Senate.[42] In 2000, he wrote his first book, *The Art of Deception*.[43] All of these achievements were in service to his longer-term goal: to establish a security consultancy. In mid-2002, he established Mitnick Security, offering his services as a social engineer for penetration testing and as an instructor for training courses to help organizations' employees recognize social engineering in action.[44]

While Headley and Mitnick made the leap from underground to professionalized, social engineering-based penetration testing, neither of them did much to formalize the field and ensure that it could be taught to others. Credit for this should go to Chris Hadnagy and Sharon Conheady. A key development in professional penetration testing is the development of a core literature. Hadnagy has built such a literature through his podcast. From its 2009 launch to the present day, Hadnagy's *Social-Engineer.org* podcast has brought on guests from law enforcement, academia, business, and the hacker underground, all with the same goal: to explore the expansive, multifaceted dimensions of social engineering as a tool for penetrating testing. And every podcast episode ends with Hadnagy posing the same question to the guests: what books do you recommend? Over the subsequent ten years, Hadnagy has collected his guests' recommendations on a blog post.[45] It's a list of more than 150 books.

The books are dominated by social scientific theories drawn predominantly from evolutionary psychology, communication, marketing, and organizational studies. The library collected by Hadnagy includes the work of mass social engineer Edward Bernays (*Propaganda* and *The Engineering of Consent*). It includes marketing and business staples, like Robert Cialdini's *Influence*, and Dale Carnegie's *How to Win Friends and Influence People*.[46] It includes guides to reading body language and emotions, including the foundational work of Paul Ekman.[47] It also has analyses on building rapport (Robin Dreeke's *It's Not All About "Me"*), thinking (Daniel Kahneman's *Thinking, Fast and Slow*), and mindfulness (multiple books by Ellen Langer).[48] And, of course, it includes a variety of books specifically focusing on social engineering, such as Johnny Long's *No Tech Hacking*, several books by Mitnick, and Hadnagy's own books.[49]

The literature helps professional social engineers understand how social engineering works for penetration testers. Whereas the phone phreaks of years past may have relied on their raw talents as they bullshitted Bell operators, contemporary social engineers have an array of social science concepts to explain how they can control other humans: reciprocity, rapport, mindfulness (and mindlessness), microexpressions, and framing. With these concepts, social engineers can describe their work in social scientific terms, further bolstering their claims to professional status, raising their esteem among clients, and reinforcing the perception that "the human is the weakest link" in security.

As a result, the professional social engineering penetration testing literature now features a stable methodology, an implementation of these theoretical concepts into practical, reportable, corporation-friendly steps. Perhaps the clearest explication of the hacker social engineering process is in Sharon Conheady's excellent book, *Social Engineering in IT Security*. The core chapters of that book are:

- Chapter 5: "Research and reconnaissance," which includes gathering OSINT, or open-source intelligence, as well as the time-tested phreak technique of trashing;
- Chapter 6: "Creating the scenario," which involves developing the pretext, including dressing the part and developing a backstory;
- Chapter 7: "Executing the social engineering test," which discusses deploying one's pretext through a variety of channels, including email, telephone, and in person; and
- Chapter 8: "Writing the social engineering report," which details how to report one's findings to the company that contracted you to test their security. This is a requirement for any professional pentester.[50]

As should be clear, we have taken these steps as guidelines for constructing our genealogy of social engineering: our chapters on trashing, pretexting, and bullshitting are roughly analogous to the research and reconnaissance, creating a scenario, and executing phases, respectively.

As for the final phase, report writing, that is the moment that penetration—the control of others, the conquest of systems—is documented. It is the culmination of a professionalized pentest, the product the client paid the professional social engineer for. As Conheady notes, the report is where the "fun" of social engineering penetration testing is transcribed into "boring" detail in presentations and written documents.[51]

But there is another method of professionally presenting results of a pentest, one that's a bit more exciting: speaking at hacker conferences. The talks given at DEF CON by Jayson Street and Susan Headley are two such examples.[52] Unlike the staid corporate report, the presentation of interpersonal hacker social engineering penetration at a conference often recaptures the fun of penetrating. Street's presentation is full of images of "security fails"—computers left unlocked, passwords left on Post-it notes, smartphones left unattended, unsecured doors. Most damning, however, are his videos of the security guards or corporate employees he's able to social engineer, using the practices of his trade to get past them and into sensitive areas.[53] For her part, Headley tells her stories about using seduction techniques to get passwords and about her habit of giving security tips to the very people she's conning. Their audiences of hackers get the vicarious thrill of seeing corporate security penetrated, again and again, while Street and Headley get credit for their abilities to penetrate.

The tension between the staid reporting Conheady details and the more ribald reporting happening at hacker conventions reflects "the tension between the subversive skills of hacking and the standardizing aims of professional certification."[54] Hacker social engineering derives its authority in part from the sort of underground, illicit activities that give it its reputation as a dangerous form of knowledge. Its professionalization is based on it being recognized by corporate and military organizations as a useful set of skills, amenable to formal reports and business hours. It takes a particular type of person—the ethical penetration tester—who can navigate this tension.

Mass Social Engineering Metaphors: Bullets

As our genealogy shows, the interpersonal hacker social engineering processes and concepts we've discussed throughout can illuminate the practices of the older, mass social engineers. Just as we can observe mass social engineering variants of trashing, pretexting, and bullshitting, we can also find precursors to the hacker logic of penetration in mass social engineering.

Like interpersonal hacker social engineering, mass social engineering is ultimately about control of people and systems. But mass social engineers reverse the relationship between people and systems. In interpersonal hacker social engineering, the social engineer penetrates the mind of his or her target as a means to penetrating the telecommunication or computer system. In mass social engineering, the media system is penetrated with the ultimate goal of penetrating the hearts and minds of human audiences. Nonetheless, mass social engineers not only talk of penetrating minds and systems, they also make similar assumptions as hacker social engineers about the nature of communication and its supposed effects.

Mass social engineering penetration is directed at mastering crowds. The idea that communication and media technologies were important to the formation and maintenance of the United States system of governance began with the founding of the country and reflected an emerging consensus among the United States' founders that media technologies, especially the newspaper, were key to turning unruly crowds into informed publics with a shared sense of understanding and opinion.[55] For example, John Adams spoke of the need for communication and transportation technologies—which were largely the same in those days—to bind the new nation together.[56] In 1787, and in response to Shay's Rebellion, Thomas Jefferson wrote that the prevention of such "interpositions of the people" required the newspaper to "penetrate the whole mass of the people" who should, he said, be sufficiently educated to read and understand them.[57] These themes would become amplified among the mass social engineers of the early twentieth century.

Penetration for crowd mastering takes on a different metaphorical meaning among mass social engineers than the later hackers. Instead of sexual conquest, the metaphor was weapons of war. An early critic of mass social engineering, Ray Stannard Baker, noted in 1906 how public relations operatives working for the railroad industry engaged in a military-style "campaign" complete with precise "shots"—that is, editorial content—fired at small newspapers around the United States.[58] Later, the WWI-era Creel Committee would adopt the bullet metaphor explicitly. American studies scholar Jonathan Auerbach notes that George Creel himself described the committee's messages as "paper bullets" and "shrapnel" in a battle for American "hearts and minds."[59] These bullets were shot through many media, from broadcast systems like newspapers and radio to more modest forms like buttons, corner speeches, and sign-boards. As Auerbach writes, Creel Committee media messages "penetrated virtually every aspect of American life."[60] The "paper bullets" penetration metaphor would go on to be a common one among analysts of wartime propaganda.[61]

These early attempts at mass social engineering were meant to exploit the lessons of Progressive social science, which taught that human behavior was fully knowable and malleable, so long as it could be penetrated with media messages. Edward Bernays, who started his career working for the Creel Committee, repeated the "paper bullet" metaphor in his 1942 analysis of US World War II propaganda.[62] He developed the penetration metaphor further in his 1947 essay on the "engineering of consent." His media effects theory was unambiguous: "communication is the key to engineering consent for social action" precisely because "the ideas conveyed by the words will become part and parcel of the people themselves."[63] The minds of the people, he argued, will be so thoroughly penetrated by the ideas suggested by the mass social engineer that the people will then act on their own accord.

Overall, the mass social engineers Doris Fleichman, Ivy Lee, George Creel, Edward Bernays, and others believed that they could impact individual perceptions and behaviors, and ultimately society through the use of scientific techniques of mass persuasion, and they often spoke of these effects in metaphors invoking the idea of penetration.[64] Just as hacker social engineering would later assume an instrumentalist model of communication in which language is a means of control through the "programming" ("neuro-linguistic" or otherwise) of the target, mass social engineers adhered to what communication theorist James Carey called the "transmission view of communication" and communication historian Christopher Simpson called "communication as domination." This instrumental model has defined US communication studies from the start, including early attempts at mass social engineering. As Simpson explains, this model of communication, which emerged out of WWI propaganda efforts like the Creel Committee and became more formally codified during WWII and the early years of the Cold War,

> concentrated on how modern technology could be used by elites to manage social change, extract political concessions, or win purchasing decisions from targeted audiences. . . . This orientation reduced the extraordinarily complex, inherently communal social process of communication to simple models based on the dynamics of transmission of persuasive—and, in the final analysis, coercive—messages.[65]

This is penetration as crowd control, paper bullets meant to manage the masses.

Media Penetration by the Numbers

Like the later professional hacker social engineers, mass social engineering was done for clients, who demanded documented results. Whether or not the crowd was penetrated by paper bullets required some form of proof. Thus, mass social engineers worked hard to prove their prowess by reporting on their successes, and the vehicle they chose was basic quantification, most commonly the counting of news stories—clips—mentioning the client.[66] Simply put, their logic was that the deeper their ideas penetrated media systems—the more mentions of their messages across various media—the more likely they had penetrated the minds of their target audiences.

Fleischman and Bernays present their work on behalf of American Tobacco in metrics. In their effort to influence fashion designers to use the color green (and thus make Lucky Strike cigarette packaging fashionable), they created the pretext of a Color Fashion Bureau.[67] Bernays and Fleischman claimed their effort to penetrate the fashion industry to be a success because of a basic metric: inquiries about green made to their pretext, the bureau.

> Just months after opening, the Color Fashion Bureau was besieged with requests for information—from 77 newspapers, 95 magazines, 29 syndicates, 301 department stores, 145 women's clubs, 175 radio stations, 83 manufacturers of furniture and home decorations, 64 interior decorators, 10 costumers, and 49 photographers and illustrators.[68]

We have such precise numbers from Fleischman and Bernays because they saw such metrics as evidence of penetration. Their work is marked by counting media clips: Bernays's *Biography of an Idea* delights in the sheer number of news stories about their efforts, and Fleischman's edited trade magazine *Contact* shared clips with subscribers as evidence of their firm's success.[69]

Metrification bolsters the mass social engineer's claims that penetration of media systems shapes the perceptions of the crowds. Ivy Lee's campaign on behalf of the Rockefellers and the coal industry in the 1910s was "a virtual avalanche of turn-of-the-century political direct mail," with tens of thousands of leaflets and booklets mailed to influential people across the United States.[70] After the avalanche of mail was sent, Lee measured the results of this by doing what we might call sentiment analysis; he hired a clipping service and an assistant to analyze news editorials, finding more than half the editorials to be favorable to the Rockefellers and the coal companies.

Of course, such crude metrification pales in comparison to the broader quantification of communication and media research that accelerated after World War II. As communication historian Christopher Simpson notes, communication researchers followed the lead of the mass social engineers in order to see mass media as a tool for social management and as a weapon in social conflict. But unlike the clip-counting practices of the mass social engineers, they proposed more complex quantitative approaches—particularly experimental and quasi-experimental effects research, random sampling, opinion surveys, and quantitative content analysis—as a means of narrowly defining communication as social management.[71] By the 1950s, an article in the academic journal *Public Opinion Quarterly* reported that the field was using a range of standardized "effectiveness studies" to gauge how deeply their messages penetrated a media system: clients "may buy a rating service which reports on the size of a television or magazine audience. [They] may study the degree of penetration which [their] message has achieved in various segments of the public. [They] may pretest the readability of [their] advertising copy."[72] Despite the variations in complexity, both the mass social engineers and the later mass communication researchers conceived of penetrating society as a matter of penetrating media systems with their preferred messages. In mass social engineering and its social scientific descendants, penetration is a numbers game. To share results with a client, point to what you can count.

Penetrating Us for Our Own Good?

Whether they seek to penetrate a building's security, a computer system, a market, or a national media system, all professional social engineers—mass or interpersonal—do their work on behalf of clients. The mass social engineers style themselves as "public relations counsels," penetrating media systems on behalf of corporations wanting to improve sales or stave off regulation, or governments wanting to improve their geopolitical positions. Professionalized pentesters do their work as consultants to or employees of corporations who want to discover possible holes in their security systems. Thus, these social engineers—like many other types of engineers—offer their technocratic talents to those in power. They penetrate us not in service to some larger ideal, but rather to meet the needs of those who pay them.

The people writing the checks out to social engineers have a lot to lose. Governments fear that some opposing political movement or government will undermine their legitimacy. Corporations dependent upon consumption fear that people will stop buying whatever they're selling. Organizations fear that their secrets will be exfiltrated and sold in black markets. Social engineers theorize all of these problems as problems of communication—specifically, problems of instrumental communication, where people are being penetrated by the wrong messages from the wrong people. This appears in their characterization of social engineering as a neutral process, a value-free "tool" that can be picked up by both "bad guys" and "good guys" alike. In fact, the bad guys, they argue, are already doing it—so the good guys simply have to.

As has been shown throughout this book, Bernays, Lee, Fleischman, and other mass social engineers stoked fears that crowds of common people would be controlled by political demagogues who would penetrate the "common man's" mind with media messages. Bernays argued,

> self-seeking men capitalized on the fact that the common man had been swayed . . . by propaganda. This powerful common man could be influenced by symbols, by words, pictures and actions. Appeals could be made to his prejudices, his loves and his hates, to his unfulfilled desires.[73]

And Lee argued,

> The crowd craves leadership. If it does not get intelligent leadership, it is going to take fallacious leadership. We know that the leadership which the mob has often received not only in this country but in other countries, unless corrected, is liable to produce disastrous consequences.[74]

In essence, Lee, Bernays, Fleischman, and other mass social engineers argued that the control and manipulation of crowds was inevitable, and in fact had already happened (predominantly by Germans and Bolsheviks).

In this sense, their observations map onto those of their critics. Critics of the emerging influence industry, including such prominent voices as John Dewey and Walter Lippmann, warned about the "threat of engineered and coercive opinion" and called for reform and revitalization of both the education system and the press.[75] The concerns expressed by Dewey and Lippmann gained increasing traction from the 1930s and into the early years of the Cold War:

> With the rise of totalitarian regimes, propaganda could no longer be innocently taken as a kind of education, shaping and organizing the intelligence of the American public; now, education was enlisted precisely to counter the power of print, radio, and cinema, all perceived as potentially threatening forms of coercion and pacification.[76]

These fears resulted in Congressional hearings and the creation of organizations like the Institute for Propaganda Analysis, both with the goal of studying the impact and spread of Nazi propaganda in the United States, as well as the creation of educational curricula to help inoculate Americans against the effects of such propaganda.[77]

However, for the mass social engineers, the solution against consent engineering was not more education, but more consent engineering. In 1947, Bernays argued that the inadequacies of Americans' education meant that leaders sometimes could not wait for people to become properly educated before making a decision.

With "pressing crises and decisions" at hand, combined with the fact that "the average American adult has only six years of schooling behind him," Bernays said, leaders had the "obligation" to use consent engineering to bring the public along to their way of thinking. Education, while still important, would not be enough on its own. "The engineering of consent will always be needed as an adjunct to, or a partner of, the educational process."[78] Consent engineers presented themselves as the ethical engineers who could translate the needs of the ruling elites and nimbly combat "fallacious leadership" of crowds through the penetration of minds qua media messages. As Lee argued in 1915, if demagogues get to direct the crowd, "why should not the same process be utilized on behalf of constructive undertakings, on behalf of ideas and principles which do not tear down but really build up?"[79] Bernays echoed Lee's argument, stating, "We must recognize the significance of modern communications not only as a highly organized mechanical web but as a potent force for social good or possible evil" and also that consent engineering practices "may be and sometimes are abused. There are demagogues not only in politics but in all branches of endeavor."[80] Evil must be engineered away, instead of ameliorated through education.

Such views persisted into the Cold War as the United States government worried about the potentially subversive effects of Soviet "psychological warfare" against Americans and others. Though, at the same time, the US government developed its own tools and techniques of political and psychological warfare for use against the Soviets, the Eastern Bloc, and third-world countries believed to be uniquely susceptible to malignant Soviet influence.[81] The penetration of Western mass media and education programs into third-world countries was a key metric for judging the success or failure of "development" and "modernization" efforts.[82] Likewise, psychological warfare techniques were seen as valuable tools for countering communist insurgencies in cases where development efforts had failed. In short, while the United States worried about the effects of propaganda at home, its ultimate position was propaganda for thee, but not for me.

For their part, professional hacker social engineers adhere to similar logics. First, they acknowledge that their brand of interpersonal social engineering is often used for malicious purposes. In his book *Social Engineering: The Science of Human Hacking*, Hadnagy tells us

> I cannot control how you use this information. You can read this book and go out and attack people and steal their information. Or you can read this book and learn how to be a defender for what is right.[83]

And Conheady's book *Social Engineering in IT Security* welcomes us "to the twisted and deceitful world of social engineering where nothing is as it seems. What you are about to read can be used for good or evil."[84]

Indeed, evil uses of hacker social engineering wisdom are, of course, already among us. As Mitnick writes in *The Art of Deception*, we need to understand "how you, your co-workers, and others in your company are being manipulated." In this vision, social engineers are already breaking our security. We need to be taught how to "stop being victims."[85] We need to become social engineers ourselves and fight for "what is right."

To aid in the fight for what is right, contemporary professional social engineers offer their services to educate the rest of us. Hadnagy is particularly keen to suggest education: "My motto," he writes, "is 'security through education.' Being educated is one of the only surefire ways to remain secure against the increasing threats of social engineering and identity theft."[86] After all, "The only true way to reduce the effect of these attacks is to know that they exist, to know how they are done, and to understand the thinking process and mentality of the people who would do such things."[87] Mitnick and Conheady use similar language.

But ultimately, Hadnagy's ideals of education—a democratic vision, redolent of the mid-twentieth-century push to educate people against propaganda—are not quite what gets put into practice. Consider his "Human Hacking Conference," an annual "educational event where you receive expert training on how to hack thoughts, actions, and the people around you. The skills and insights you gain from attending the HHC benefit you both personally and professionally."[88] Rather than train broad sectors of society on how social engineering works, Hadnagy's conference, and the books by professional social engineers, are aimed at reproducing the field of professional social engineering by educating the next generation of pentesters. The newly minted professional social engineers can then carry on the legacy of offering ethical penetrating services to test the security of large organizations, providing reports to those organizations on how to improve their security. Much as the mass social engineers offered their consent engineering approach as an antidote to malicious propaganda, professional social engineers offer their services to combat malicious social engineers.

Conclusion

What American studies scholar Jonathan Auerbach argues of mass social engineering is equally true of interpersonal hacker social engineering: their penetrative powers are "at once part of the problem as well as a potential solution—a way to control and direct an uncertain, disparate citizenry, but also possibly to mobilize and guide it toward a greater common good."[89] The deployment of ethical hacker social engineers is an attempt to "appropriate the technical authority and mystique of hackers . . . without the stigma of the popular association of hackers with criminal activity."[90] If malicious hacker social engineers are controlling your employees in order to gain access to your corporate systems, then the best defense is to hire someone to hack, pwn, own, and penetrate those same employees. If malicious mass social engineers are hitting the hapless masses with paper bullets, then in this way of thinking, the only viable response to a bad guy with a message gun is to hire a good guy with a message gun loaded with more and better paper bullets.

Thus, what this analysis of penetration teaches us is that those in power are the ones in a position to wield the trashing, pretexting, and bullshitting capabilities of social engineers. Whether social engineering is intended to subdue crowds or control individuals, it is most often in the service of those with the resources to hire social engineers. These are often the selfsame people who distinguish between good social engineering and bad. And thanks to new developments in media systems—specifically, the advent of corporate social media—social engineering is available in a new form: a masspersonal form. We turn to that next.

III
Masspersonal Social Engineering

7
Contemporary Masspersonal Social Engineering

> In this new war, the American voter became a target of confusion, manipulation, and deception. Truth was replaced by alternative narratives and virtual realities.
> —Christopher Wylie[1]

In this chapter, we return to the cases of Russia's and Cambridge Analytica's attempts at election interference and manipulation to help us elaborate on the concept of masspersonal social engineering. By focusing on 2016, we do not mean to provide a definitive account of those events, but rather to use them as a window into the emerging intersections and blurring of boundaries between the mass and interpersonal forms of social engineering discussed in the last several chapters. Though we will also discuss more recent examples of the practices of masspersonal social engineering, the 2016 case remains the most well documented. Because of this, the 2016 Russian operation and Cambridge Analytica will be our primary focus in this chapter.

So far, we have taken up the provocative phrase, "social engineering," pursuing its genealogy and tracing its practices. We argued that social engineering has two distinct meanings in American thought. One is the form we have been calling "mass social engineering," a mentality that dates back to the Progressive Era of the early twentieth century and sees society as comprised of unruly crowds and their elite betters who should manage those crowds through mass communication. The proponents of this view—Ivy Lee, Doris Fleischman, and Edward Bernays—argued that scientific analysis of society reveals underlying laws that, once understood, allow elites to implement programs that can "engineer the consent" of crowds with carefully crafted messages. Their vision was of the crowd as something in need of, as Bernays puts it, "adjusting" to the dictates of corporate or government elites.[2] Doing so would increase the efficiency of consumption and production, ease social conflicts, and provide support for government programs, including war-making. However, this ambitious program of social adjustment via mass communication was largely delegitimated by the mid-twentieth century.

In the mid-1970s, just as the mass social engineering idea was undermined, a second meaning of social engineering emerged in American thinking: a new, interpersonal form of con artistry developed by the phone phreaks and, later, computer security hackers. The hacker social engineers tend to target individuals, seeking to manipulate them into revealing sensitive information or giving access to information systems. This interpersonal hacker form of social engineering was unlike its predecessor in that it did not start with a science-to-implementation vision—indeed, this version of social engineering was a polite way of saying "bullshitting"—but eventually, as computer security hacking became a professionalized field in the 2000s, so too did hacker social engineering, using social science to explain itself to the broader world.

As we argued in the introduction, both the Russian operation and Cambridge Analytica involved both mass and interpersonal social engineering. Russia's efforts were aimed at the masses insofar as they targeted populations. But the Russians also used interpersonal con artistry to achieve their goals, as in the case of the spear phishing attack on John Podesta or Internet Research Agency trolls interacting with Americans via Instagram. Cambridge Analytica sought to affect national elections through the use of big data about voters. But Cambridge Analytica's sales pitch to political campaigns centered on its ability to "microtarget" messages to individuals via social media. At the very least, both the Russian interference campaign and Cambridge Analytica demonstrated the same ambitions as the older mass social engineers, but, as we will discuss, they also used the more targeted interpersonal techniques developed by hackers and phone phreaks.

This brings us to our larger argument: in their use of both mass and interpersonal forms of social engineering, both the Russian effort and Cambridge Analytica represent something new, something in addition to or synthetic of the previous forms of social engineering. Mass social engineering was developed in the context of mass communication. Hacker social engineering privileged interpersonal communication. Russia's operation and Cambridge Analytica relied upon a convergent, fluid mix of both, a form we call *masspersonal social engineering*, which is enabled by the unique affordances of the internet and social media platforms and brings together the tools and techniques of hackers and propagandists, interpersonal and mass communication. While the mass social engineers used radio and television, and the phone phreaks and hackers used phones, emails, or even in-person engagements, masspersonal social engineers use corporate social media: Facebook, Instagram, and Twitter, with their peculiar mix of one-to-one and one-to-many capabilities.

Communication theorists O'Sullivan and Carr argue that masspersonal communication occurs when "individuals use conventional mass communication channels for interpersonal communication, individuals use conventional interpersonal communication channels for mass communication, and individuals engage in mass communication and interpersonal communication simultaneously."[3] That is, rather than focus on the channels as determining the type of communication, O'Sullivan and Carr suggest we attend to the "interactional goals" of communicators, who might reinvent what appears to be a mass communication technology for interpersonal communication, or vice versa. Both the Russian operation and Cambridge Analytica had similar interactional goals: affecting democratic deliberation and voting during the 2016 US presidential election. They then used the masspersonal affordances of corporate social media in pursuit of that goal.

The Social Engineering Process

To further elaborate our claim that the Russian operations and Cambridge Analytica are exemplary forms of masspersonal social engineering, we first want to attend to the social engineering process in general. Taking up the two meanings of social engineering, in the preceding pages we've laid out the general process by which social engineering (in all senses) typically works.

Overall, to judge whether other events are instances of masspersonal social engineering, our heuristic is, if the communicative practice:

- relies upon what we're calling trashing, or intense data-gathering about targets;
- utilizes pretexts, or roles that are recognizable to others yet are based on deception and obscure the identity of the messenger;
- involves the conversational practices of bullshitting—a truth-indifferent mix of friendliness, deception, and accuracy;
- has the goal of penetrating something—a mind, a machine, a news cycle, an election; and

- uses a mix of personalized, targeted messages and mass media messages,

then the practice is masspersonal social engineering. Conversely, a communicative practice that has some but not all of these elements is probably not.[4]

Such a heuristic helps avoid the common trap of thinking that every manipulative communicative act is social engineering. The presence of any individual practice—trashing or bullshitting, for example—does not necessarily mean that masspersonal social engineering is happening. (Otherwise, acts like advertising or political speeches would be classified as masspersonal social engineering.) Masspersonal social engineering happens when all these elements are present and the social engineer fluidly shifts from targeting an individual to targeting crowds and back again.

These broad steps have been amplified in the past decade—a fact made very clear by both the Russian Internet Research Agency and Cambridge Analytica. Because these cases are explosively controversial and thus are well documented, they allow us to explain the masspersonal social engineering process in detail.

Masspersonal Social Engineering Cases: Russia and Cambridge Analytica

Trashing

Let's begin with Russian trashing.

In order to get started, the Russian operation had to wade through America's digital dumpster, delving into massive amounts of data in order to find valuable pieces of information.[5] For example, the phishing attack on Clinton campaign chair John Podesta was not initially a targeted attack against a high-level Clinton campaign staffer. Rather, the Russians had to sort through a pile of digital detritus. As *AP News* reports,

> One of the first people targeted was Rahul Sreenivasan, who had worked as a Clinton organizer in Texas in 2008—his first paid job in politics. Sreenivasan, now a legislative staffer in Austin, was dumbfounded when told by the AP that hackers had tried to break into his 2008 email—an address he said had been dead for nearly a decade.
>
> "They probably crawled the internet for this stuff," he said.
>
> Almost everyone else targeted in the initial wave was, like Sreenivasan, a 2008 staffer whose defunct email address had somehow lingered online.[6]

Sreenivasan is of course forgiven for forgetting all about an old email address of his. Like many of us might, he threw it away when that part of his career was over. It took several waves of working through similar discarded email addresses that were known to be registered with the hillaryclinton.com domain before Russia's Fancy Bear team found active email accounts, narrowed their list of targets, and eventually sent a malicious spoofed link to an unsuspecting John Podesta. Like the trash bins behind the phone company, the internet itself contained out-of-date and discarded information that had to be sifted through to glean enough to refine an attack.

As for the Internet Research Agency and its social media trolls, the US Department of Justice indictment against twelve of its employees claims that their trashing efforts began as early as May 2014 when the IRA began to discuss interfering in the upcoming US presidential election and started "monitor[ing] US social media accounts and other sources of information about the 2016 US presidential election."[7] More specifically, the indictment explains that the IRA

> began to track and study groups on US social media sites dedicated to US politics and social issues. In order to gauge the performance of various groups on social media sites, [the IRA] tracked certain metrics like the group's size, the frequency of content placed on the group, and the level of audience engagement with that content, such as the average number of comments or responses to a post.[8]

Presumably, this information was analyzed by the Internet Research Agency's dedicated "data analysis department."[9] Such a department joins the myriad others around the world picking through the waste product of contemporary digital capitalism: the traces left behind as we engage with digital media. The conversion of such traces into valuable insights is a trash-to-treasure process, gleaning valuable information out of a massive collection of data.[10] Such efforts can be done for a range of reasons, including gaining sociological insights into human behaviors, epidemiology, marketing, or, of course, social engineering.

As for Cambridge Analytica, perhaps no company in the world better exemplifies the perils of the digital dumpster—with the possible exception of Facebook.[11] The most controversial aspect of the Cambridge Analytica saga was its use of unethically obtained—and, in some jurisdictions, possibly illegally obtained—Facebook data. As the *New York Times* reported in 2018:

> As the upstart voter-profiling company Cambridge Analytica prepared to wade into the 2014 American midterm elections, it had a problem.
> The firm had secured a $15 million investment from Robert Mercer, the wealthy Republican donor, and wooed his political adviser, Stephen K. Bannon, with the promise of tools that could identify the personalities of American voters and influence their behavior. But it did not have the data to make its new products work.[12]

So, Cambridge Analytica went trashing. It acquired data from the Cambridge University researcher Aleksandr Kogan, who had gathered it from people who used the "thisisyourdigitallife" personality test Facebook app in 2014.[13] And Cambridge Analytica also took the data of the *friends* who used that app. As a result, "only a tiny fraction of the users [of the app] had agreed to release their information to a third party."[14] Moreover, the data collected for the personality app was intended for academic, not commercial use. But, like a stack of documents in a dumpster outside a telco office, the data were just ripe for the taking. As the Cambridge Analytica whistleblower Christopher Wylie put it, "Where [the data] came from, who said we could have it—we weren't really asking."[15]

There were additional apps being used to profile Facebook users and gather their friends' data, apps with names like "Music Walrus" and the "Sex Compass."[16] Wylie notes that

> Cambridge Analytica [also] began testing innocuous-looking browser extensions, such as calculators and calendars, that pulled access to the user's Facebook session cookies, which in turn allowed the company to log in to Facebook as the target user to harvest their data and that of their friends.[17]

In addition, the company purchased more data from firms such as Experian, Acxiom, Magellan, and L2.[18] Apps, browser extensions, and data markets: the digital dumpster is vast, and Cambridge Analytica was happily wallowing in it.

Like any other big data analysis operation, Cambridge Analytica had to clean the data, echoing the older phone phreak act of sorting through the trash. Cambridge Analytica whistleblower Brittany Kaiser calls this "hygiene-ing . . . , the process by which data engineers match new data to old and fix errors. . . . The cleaner the data, the more accurate the algorithms, and hence, the better the predictability."[19] Kaiser's highlighting of so-called "hygiene-ing" reminds us of "the complicated and material processes involved in treating data and turning it into recyclable resources for new forms of knowledge production. With datafication we thus find ourselves in the waste incinerator, not the gold mine."[20] Like the phreaks and hackers who would jump in dumpsters and grab bags for later sorting, Cambridge Analytica gathered massive gluts of data but understood that they had to sort it to find the valuable information they needed for their psychographic profiles.

Pretexting

As multiple analyses show, Russia's Internet Research Agency engaged in pretexting, masquerading in a wide range of roles before, during, and after the 2016 election. These pretexts included roles such as Black activists and Texas conservatives. These pretexts are not mysterious or magical; they reflect preexisting roles that were highly recognizable in the years immediately preceding the election, apparent to the Russians after they had completed their trashing efforts. As former US National Security Council member Fiona Hill explains, "The Russian state does not meddle directly. It delegates to proxies, who amplify our divisions and exploit our political polarization."[21]

The political polarizations the Russians exploited included divisions over racial experiences, gun-regulation stances, and immigration reform. As the 2019 State of Black America Report notes, during the election, "Russian propagandists specifically targeted African Americans through a wide-reaching influence campaign. Their tactics included posing as legitimate activist groups, eroding trust in democratic institutions, and spreading disinformation."[22] Likewise, a US Senate-commissioned report explains how

> IRA posts [also] tended to mimic conservative views against gun control and for increased regulation of immigrants. In some cases, terms such as "parasites" were used to reference immigrants and others expressed some tolerance of extremist views.[23]

Hence, the IRA relied upon existing and recognizable divisions and social stereotypes as a set of possible online pretexts.

These pretexts were deployed in a range of ways. One approach was to use socialbots, or automated social media profiles that appear to be human.[24] Socialbots are "force multipliers" that "leverage machine learning and artificial intelligence to conduct targeted and timely information transactions at scale while leaving critical nuanced dialog to human operators."[25] The IRA used socialbots to subtly shape online discourse on a mass scale. As the Mueller Report documents, Russians used bots in Twitter extensively:

> In January 2018, Twitter publicly identified 3,814 Twitter accounts associated with the IRA. According to Twitter, in the ten weeks before the 2016 US presidential election, these accounts posted

approximately 175,993 tweets, "approximately 8.4% of which were election-related."[26]

While its analysis focuses on an election in Sweden and not the 2016 US presidential election, a report sponsored by the US Office of Naval Intelligence includes data about the extent of Russian bot-like (i.e., very probably, but not confirmed, socialbot) behavior on Twitter during the run-up to the Swedish election, showing the extent to which Russian bots insinuated themselves into online debates.[27] Researchers focusing on the US context find similar dynamics.[28]

But the Internet Research Agency didn't just use bots. It also utilized human-controlled social media profiles as pretexts. "While media narratives around the Russian/IRA Twitter activity have often focused on automation and bots, the agency ran human-operated precision personas that roughly mapped to the same Black, Left, and Right clusters observed on Facebook and Instagram."[29] While socialbots are useful for relatively crude pretexts that automatically make social media posts, human-controlled social media accounts can be more flexible, allowing for more creative bullshitting.

As pretexts, these operations mapped onto and amplified in-group and out-group dynamics: "Facebook and Instagram were used to develop deeper relationships, to create a collection of substantive cultural media pages dedicated to continual reinforcement of in-group and out-group ideals for targeted audiences."[30] The success of these pretexts is marked by how well they were recognized by actual Americans, who in turn helped legitimate the pretext: "The goal of working with real Americans is to eliminate the detection and exposure risk of inauthentic personas."[31] While much of the commentary on the Russian operation decries it as "fake" or "inauthentic," what the IRA pretexts reveal is how careful the Russians were in emulating the dynamics of American culture.

Cambridge Analytica also relied heavily on pretexting. The first and most well-known layer of Cambridge Analytica pretexting is closely tied to the company's trashing activities. The company's cooperation with Cambridge University's Dr. Aleksandr Kogan was first aimed at helping promote the campaigns of Ted Cruz and then Donald Trump in their respective 2016 presidential bids. But this trashing was enabled by a pretext. Kogan used Amazon's crowdsourcing platform, Mechanical Turk, to recruit and pay participants to take a personality quiz that also gave him access to their Facebook profiles and the profiles of their friends. The personality data and Facebook profile data were then matched and used to create a predictive model deployed to drive the segmentation, targeting, and personalization of political messaging on behalf of the company's clients. All of this was done under the pretext of "academic research" and participants were told that their data would be anonymous and kept confidential. What they did not know was that Kogan was acting not in his capacity as an academic, but rather, was being paid by Cambridge Analytica to collect this data under the name of Global Sciences Research, a company Kogan established for precisely this purpose. Not only was this data collection not primarily about academic research, but Kogan's own identity as a Cambridge academic researcher had other, undisclosed layers. The *Guardian* newspaper reported that Kogan "had previously unreported links to a Russian university and took Russian grants for research" and that the research used as pretext for data collection was actually a replication of work done by two colleagues meant to cut them and the university out of any proceeds for contracted work with Cambridge Analytica.[32]

Beyond Kogan's data gathering, the various Facebook apps Cambridge Analytica used for the same purpose—the Music Walrus and the Sex Compass—were not presented as data-collection efforts but as entertaining diversions. Moreover, they were not presented as belonging to Cambridge Analytica. Speaking to CBS reporters in 2018, former Cambridge Analytica employee Brittany Kaiser

> showed us a 2015 note from a data scientist behind a quiz called "Music Walrus." In the all-company email, he asked colleagues

to share it with family and friends but added, "Please DO NOT MENTION THAT IT IS [OUR CREATION]. Say that a 'friend of yours' made it (which is true . . . no?)"[33]

These apps were presented as harmless entertainment for friends and family—a fine pretext for data gathering.

But the company is reported to have used more nefarious pretexts than misleading apps. In 2018, based on a series of secret recordings of Cambridge Analytica executives, the UK's *Channel 4* reported that the company had also deployed a range of "dirty tricks" for collecting damaging information about its clients' political opponents. Alexander Nix, then the company's CEO, described these tactics on a secret recording, saying, "We have two projects at the moment, which involve doing deep, deep depth [sic] research on the opposition and providing source . . . really damaging source material, that we can decide how to deploy in the course of the campaign."[34] These projects included sending researchers into foreign countries, like the United States, with the pretexts of being tourists or students. More disturbing still, it included the use of former British and Israeli spies to aid the company in hiring prostitutes to set honey traps, arrange fake bribery stings, and more.[35]

Finally, we might go a step further and say that Cambridge Analytica is itself a pretext that is built on top of, or operates within, pretextual platforms. Using data to segment and target individuals and small groups with customized messages is not so out of the ordinary for an online marketing company these days. As Nix told an audience at the Online Marketing Rockstars in 2017, the work they did for the Donald Trump campaign is representative of the growing importance of behavioral science, data analytics, and addressable ad tech for the world of online marketing.[36] He is not wrong. But we now know that Cambridge Analytica was no run-of-the-mill online marketing firm. It was a firm focused on politics—right-wing politics in particular—with ties to right-wing American political figures such as Steve Bannon and Robert Mercer. What's more, it was not just a digital political marketing firm but also a subsidiary of the SCL Group (formerly Strategic Communication Laboratories). Once again, this sounds like an innocuous public relations or marketing firm, but it was really a defense contractor reportedly engaged in "cyberwarfare for elections."[37]

In short, in almost every way, Cambridge Analytica was more, or other, than it portrayed itself to be. And this pretexting was by design. As Nix told an undercover reporter, "We're used to operating through different vehicles, in the shadows." The managing director of Cambridge Analytica's political division, Mark Turnbull, explained further that the company's activity, "has to happen without anyone thinking it's propaganda, because the moment you think 'that's propaganda' the next question is: 'Who's put that out?'" To cover their tracks, he said, "It may be that we have to contract under a different name . . . a different entity, with a different name, so that no record exists with our name attached to this at all."[38]

Bullshitting

Recall that bullshitting is the core activity of social engineering. Bullshitting occurs when the pretext goes live, when the social engineer engages with the target and has to develop rapport, carefully suggest messages and ideas, and not blow their cover. At this point, the social engineer has to be nimble, building a relationship with the target within the framework established by the pretext. This requires a skillful blend of deception, friendliness, and accuracy, all with the goal of penetrating the target. If the truth appears, it's beside the point—the penetration is what matters.

That the Russian bullshit included deception should be obvious: whether deployed as bots or human-controlled accounts, the Internet Research Agency operatives were not actually Black Lives Matter activists or gun-toting Texans advocating for the Second Amendment. Vlad from Vladivostok was decidedly not Dewayne from Detroit or Harold from Houston. However, analysis that stops at this point and simply decries the IRA efforts as completely fake misses much of what the IRA was up to.

Notably, decrying the IRA as fake ignores the sheer friendliness of the IRA posts—the seemingly earnest desire to help build Black activists or Texas conservatives into an online community. The Mueller Report noted that only 8.4 percent of the Tweets that Twitter identified as coming from IRA accounts had to do with the election. Likewise, another analysis of a random sample of IRA Tweets found that the majority of them (52.6 percent) had "no clear or overt connection to the IRA agenda building activity"—that is, no connection to electoral politics.[39] This begs the question: what is the purpose of this large majority of social media posts? Rather than bracketing these apolitical posts off, we have to see them in the larger context of Internet Research Agency activities. That is, we should consider them as the "sociable" element of the bullshitting and then examine how the IRA mixed in deception and accuracy.

A key visualization of the sociable ingredient working alongside the other aspects of bullshit can be seen in a report which includes screenshots of Instagram and Facebook image macro-style memes that were traced to IRA accounts. While there are pointed political critiques, many of the posts are benign, even inspirational. Most telling is a collection of Texas-oriented image memes from an IRA-run Facebook group called "Heart of Texas." The images mix together anti-immigrant and Texas secessionist (#Texit) rhetoric with a collection of pictures of Texas sunsets and wildflowers.[40]

Very often, analysts who look at these types of posts categorize them as "not directly political."[41] However, when we place posts such as these in the broader context of social engineering, these posts should be seen as the sociable side of a larger bullshitting campaign. In the Heart of Texas Facebook group, a steady stream of Texas sunsets and wildflowers is a fine way to amplify Texas pride, which in turn is articulated with anti-immigrant sentiment or bogus statistics about the economic success of Texas should it secede from the rest of the United States. Likewise, the stream of Black pride images presented by Russian operatives were articulated with narratives about the hopelessness of voting in a white supremacist society. Taking these posts out of context and analyzing them as strictly "not directly political" distracts us from the more disturbing goals of Internet Research Agency bullshitting: to discourage Black Americans from taking part in electoral politics or stoking #Texit secessionist politics among Texans.

As for the other element of bullshit, accuracy, such sociability is part and parcel of what the philosopher Harry Frankfurt called the "exquisitely sophisticated craftsmen who . . . dedicate themselves tirelessly to getting every [bullshit] word and image they produce exactly right."[42] This is reflected by the fact that the majority of Russian-controlled accounts were "engaged in mimicking the kinds of interests and values coherent with the social identities that they were 'spoofing,' and then occasionally they would message avowedly political content."[43] The Internet Research Agency social media campaign was designed to create networks of "authentic" and "inauthentic" accounts.[44] This required the accuracy aspect of bullshit, using language and images in a way that would lead real Americans to recognize the IRA pretext as legitimate. Doing so created a mix of actual accounts and vetted ones, so that

> An individual who followed or liked one of the Black-community-targeted IRA Pages would have been exposed to content from dozens more, as well as carefully curated authentic Black media content that was ideologically or thematically aligned with the Internet Research Agency messaging.[45]

Once someone engaged with this networked mixture of real and pretext people, the IRA would interpersonally interact with them, with personas that "were spontaneous and responsive, engaging with real users (famous influencers and media, as well as regular people), participating in real-time conversations, creating polls, and playing hashtag games. These personas developed relationships with American citizens."[46]

Here, the desire to get American culture right is part of the larger bullshitting process. Successful play at a hashtag game, where social media users post responses to prompts, often in a playful, culturally adept manner, means the IRA got their cultural references right. With that achieved, and with a bit more sociability, the Russian bullshitter could post just about anything and have it seem at least somewhat credible. After all, it came from a persona who was vetted as a real American.

None of this is to deny the fact that the IRA was, at the end of the day, indifferent towards the truth. One example is "narrative switching," or the sometimes sudden shifts in topics even on the same accounts. "The IRA not only switched from banal to pro-Russian views but also switched abruptly between different political positions according to current Russian operational priorities, or even just to create confusion."[47] In other words, the IRA tried lots of different topics over time, even with the same account, in an attempt to see what bullshit would work best. And when Internet Research Agency employees were engaged in staged "debates" among themselves, played out in news comment fields, they used a very basic heuristic: "We don't talk, because we can see for ourselves what the others are writing, but in fact you don't even have to really read it, because it's all nonsense. . . . And how you write doesn't matter; you can praise or scold."[48]

As for Cambridge Analytica, their use of bullshit is more intricate than the Russian effort. Perhaps this is in part due to the company employees' deeper understanding of American culture. It is also due to the fact that Cambridge Analytica's work blurred the boundaries between pretexting, trashing, bullshitting, and penetrating in its iteration of masspersonal social engineering.

We do not know as much as we would like about the actual messaging and interactions between Cambridge Analytica and its targets from public reporting. Part of the reason, of course, is that the messages were targeted as "dark ads" visible only to certain narrow segments of people via social media, and social media corporations, including Facebook, are not providing details about these ads. If you were not part of these segments—the vast majority of us were not—then you very likely never saw any of Cambridge Analytica's messages.[49] But the ads and campaigns we know about are quite impressive examples of bullshit. One of the few we can still access is an ad the company ran in *Politico*. Cambridge Analytica employee Molly Schweickert referred to this ad as a kinder, friendlier alternative to the traditional, blanket 30 second commercial: the ad was "nice-looking and pretty."[50] The ad is indeed user-friendly: a few clicks or taps provides startling financial details about the Clinton Foundation, relying heavily on highlighted quotes from news articles and complete with links to the original sources. However, multiple fact-checking organizations have repeatedly found Trump camp claims about the Clinton Foundation to be false.[51] But, as Cambridge Analytica's Mark Turnbull explains, "it's no good fighting an election campaign on the facts, because actually it's all about emotions."[52]

Thanks to *Channel 4*'s sting, we also know that Cambridge Analytica executives claimed to have created the "Defeat CrOOked Hillary" slogan with the "OO" looking like handcuffs.[53] Whistleblower Christopher Wylie reports that "Build the Wall" and "Drain the Swamp" "were the exact phrases Cambridge Analytica had tested and included in reports sent to [Steve] Bannon well before Trump announced" his candidacy.[54] In each case, of course, Cambridge Analytica was indifferent to the truth behind such messages. Whether Hillary Clinton had actually committed crimes for which she should be locked up, whether a border wall made any sense as a response to immigration, and whether Trump was the man to drain the swamp or, rather, a swamp dweller himself, were all beside the point, which was merely to achieve results for the client. Wylie goes so far as to characterize the Cambridge Analytica practices as "a non-kinetic weapon designed for scaled perspecticide—the active deconstruction and manipulation of popular perception."[55] Other techniques used by Cambridge Analytica to engage in "perspecticide" include "fake pages on Facebook and other platforms that looked like real forums, groups, and news sources," such as locally focused "right-wing pages with vague names like Smith County Patriots or I Love My Country."[56]

Just because Cambridge Analytica sought to destroy perspective doesn't mean they weren't friendly about it. Of course, "friendly" can refer to being kind or amicable. This is certainly a part of bullshitting that we have discussed in previous chapters. But the belligerent calls to arrest Clinton indicate a different sort of amicability, a heightened sense of us-versus-them. Cambridge Analytica content was meant to be "friendly" to its target audience by being resonant or sharing an affinity with the targets' personalities and worldviews as inferred by Cambridge Analytica's profiling of them. That is, Cambridge Analytica's bullshit was created and targeted deliberately at those deemed to be favorably disposed to, or inclined to support, the messages being sent. Wylie describes how this was done by "targeting people with specific psychological vulnerabilities" and then "priming" them with "bits of salient information" in an effort to affect how they feel, think, and act.[57] The specific "psychological vulnerabilities" that Cambridge Analytica targeted were "people with neurotic or conspiratorial predispositions."[58] We might say, then, that Cambridge Analytica's particular brand of bullshit was neurotic- or conspiracy-"friendly."

But as Wylie's point about "perspecticide" indicates, Cambridge Analytica's overall goal was grander than targeting so-called neurotics. It sought to spread the very indifference to truth that lay at the heart of its interactions with its targets. For example, Wylie argues that Steve Bannon was using Cambridge Analytica as a psychological warfare tool in his alt-right, culture war insurgency, deliberately targeting Americans to cause "confusion, manipulation, and deception. Truth was replaced," Wylie said, "by alternative narratives and virtual realities."[59] For Bannon, Wylie explained, the remaking of society starts with deliberately causing "chaos and disruption." The undermining of truth and its replacement with nonsense was a key tool towards that end.[60] In short, the ultimate goal of Cambridge Analytica's bullshit was not merely to promote Trump directly, but also to undermine the very notions of truth and certainty, to instead promote consistent uncertainty to the neurotic and conspiratorial among us and thereby cause chaos that could be exploited by Trump and the alt-right. Of course, such tactics dovetail with the production of "unreality" through a "firehose of falsehood" found in the Russian iteration of masspersonal bullshitting.[61]

Fear appeals may not seem entirely friendly and kind, but Cambridge Analytica employees were also quite good at conversation. In a throwback to phone phreak bullshitting, Wylie recalls that he, then-CEO Alexander Nix, Bannon, and others made calls to individual Americans using the data they had collected. Using the pretext of being survey researchers from the University of Cambridge, the men would proceed to ask particular questions to which they already knew the answers based on the profiles they had compiled about their victims, including jovial banter about their favorite TV shows and religious beliefs. The calls served as a kind of informal validation of the data collected in the profiles but also, as Wylie recounts, good fun too. He writes that the phone calls pleased Bannon: "When I looked over at Bannon, he had a huge grin on his face."[62]

If there is a captain of Cambridge Analytica's Bullshit Team, it has to be Nix. Reporters and academics analyzing his sales pitches for Cambridge Analytica discuss the difficulty in picking out truth from deception. As the UK's *Channel 4* report puts it, we are never sure if Nix is offering "true backdrop, or just bravado."[63] But over a friendly dinner with a prospective client, Nix offers up a theory of bullshit, arguing that the messages that Cambridge Analytica produces "don't necessarily need to be true, as long as they're believed."[64] Such can be said not only of Cambridge Analytica's messages but also of the company's claims about its own effectiveness. One of Nix's favorite examples of effective communication is pure bullshit. The wealthy Briton offers two messages to keep people off of a private beach. One is a sign that says, "Private Property Beyond This Point." In contrast to this merely informational sign, Nix offers one meant to change behaviors: "Warning: Shark Sighted—Keep Out."[65] Are there sharks in the restricted part of the ocean? Who knows? Nix is indifferent. The point is to change behaviors of would-be swimmers, not to inform them.

Perhaps the most impressive example of Nix's bullshit came when he wooed Bannon. Wylie describes how Nix's usual approach to wooing clients by taking them to posh restaurants and private clubs in London was not working with Bannon. Nix realized "that everyone, including Bannon, suffers the yearning of an unfulfilled secret self." Bannon thought of himself not merely as a website editor or political advisor but as a philosopher and an intellectual of a new version of right-wing politics. "To win him over," Wylie writes, "Nix would need to help him achieve his fantasy of becoming a thinker of big thoughts." So instead of meeting Bannon at Cambridge Analytica's London office during his visits, Nix proposed that they meet instead at the firm's office just down the road from Cambridge University, whose "Gothic halls and sprawling greens" were more appealing to Bannon than the "fancy clubs, expensive wines, and fat cigars" that Nix usually trotted out for would-be clients in London. The only problem? Cambridge Analytica did not have an office in Cambridge. So, Nix rented out a space and created one. Whenever Bannon came to town, a contingent of Cambridge Analytica workers—along with some stand-ins hired just for the day—would relocate to Cambridge Analytica's Cambridge office where they would put on a great show for Bannon, who didn't seem to notice that some of the computers weren't even plugged in and some of the stand-in employees didn't speak English. This "Potemkin Site," Wylie concludes,

> perfectly encapsulated the heart and soul of Cambridge Analytica, which perfected the art of showing people what they want to see, whether real or not, to mold their behavior—a strategy that was so effective, even a man like Steve Bannon could be fooled by someone like Alexander Nix.[66]

Penetrating

Investigative reporting on Russia's IRA revealed something quite boring, but extremely important: working there was a job. And like many jobs, there are performance metrics. Such metrics reveal short-term and long-term penetration goals.

In the short term, the IRA employees had to achieve certain performance metrics: manage a number of accounts, gain social media followers, post a certain number of comments to media outlets. Like the "game" that pickup artists discuss, where the goal is to get in front of as many attractive women as possible, the IRA employees had to get their messages out in a specified number of posts. A report discussing work conditions at the Internet Research Agency reveals that employees "were required to make 135 comments per twelve-hour shift, working in internet forums and that they would be provided with five keywords to feature in all posts to encourage search engine pickup."[67] Some IRA employees were responsible for a handful of Facebook groups, with the goal of gaining five hundred followers in a month. Others had to maintain ten Twitter accounts and manage those accounts' followers, sometimes numbering in the tens of thousands.

Beyond increasing followers on social media, a marker of penetration is getting people to take actions. The US Department of Justice indictment against the Internet Research Agency details how the IRA successfully convinced a number of individuals and groups who supported the Trump campaign in Florida, including several Trump campaign officials, to take action in the real world, such as planning and holding political rallies.[68] In addition, the US Senate Select Committee on Intelligence found that the Russians were also able to convince Black Americans to set up self-defense courses:

> IRA operatives also spearheaded and funded a self-defense program that entailed African-American trainers being paid to teach courses in their communities. As part of this operation, an African-American activist was paid roughly $700 to teach 12 self-defense classes in a local park under the auspices of the IRA-administered "BlackFist" Facebook page.[69]

Convincing Americans to take real world actions continued after the election. As a *Politico* story reports, IRA operatives were able to convince anti-Trump activists to give them control of their Facebook events pages in exchange for sponsoring a live protest against Trump after he took office.[70]

But, of course, beyond gaining more followers or convincing people to take actions, the long-term goal of the Russian operation was to affect the democratic process in the United States. In 2016, that meant undermining the Clinton campaign (as well as any other anti-Russia political candidates).[71] The debate about the extent of the Russian operation's effects is still ongoing (and we will address the question of effects in the next chapter), but at the very least, US intelligence officials found that the Russian operation had *some* effect. "We assess the Russian intelligence services would have seen their election influence campaign as at least a qualified success because of their perceived ability to impact public discussion."[72] The extent of the qualification will take years to figure out, but insofar as the IRA held up a mirror to American society, it succeeded in casting doubt on American electoral processes, the motives of government officials, and on the possibilities of easing the sometimes violent tensions in the country. At the very least, if the goal of Russian social engineering is to undermine truth, knowledge, expertise, institutions, and above all to create a chaotic bullshit-storm that can be exploited later to promote Russia interests, then the penetration was successful, with the consequences continuing to play out to this day.

As for Cambridge Analytica, its most notorious claim was about its ability to "microtarget" individuals with societal-level implications. Such targeting was aided by what the company claimed was a deep penetration into the minds of audience members, a method of audience segmentation called psychographics. As Alexander Nix put it, Cambridge Analytica's penetrating desire was "to probe an altogether deeper motivation" than simple information-seeking.[73] Rather than rational debate, its goal was to trigger emotions, particularly fear. "Our job," claimed Nix's colleague Mark Turnbull, "is to drop the bucket further down the well than anybody else, to understand what are those really deep-seated, underlying fears [and] concerns."[74] The data the company gathered through its various trashing efforts were analyzed for the purposes of psychographic targeting, dividing people up by shared qualities such as openness or neuroticism.[75]

Such profiling, Cambridge Analytica sales staff claimed, leads to high precision messaging. As the whistleblower Brittany Kaiser recalls, her pitch to potential clients ended: "what Cambridge Analytica offers is the right message for the right target audience from the right source on the right channel at the right time. And *that's* how you win."[76] As she recalls, Nix specified these practices in their pitch to a Nigerian political party, promising that Cambridge Analytica could "address individual villages or apartment blocks, even zoom right down to particular people" with political messages designed specifically for the intended targets.[77]

Like the mass social engineers before it, Cambridge Analytica used bullet metaphors to explain its penetration, describing its data scientists as an "in-house army" armed with an "arsenal of data" that allowed them to "laser in" on their exact "targets."[78] The paper bullets of yesteryear are replaced, it seems, with focused light—perhaps more fitting for our time of fiber-optic networks and LED screens. However, Nix also often used the male, sexual conquest metaphor. According to Wylie, Nix described everything in terms of sexual conquest: in the early stages of negotiations, the two sides were "feeling each other up" or "slipping in a finger." When a deal closed, he'd exclaim, "now we're fucking!"[79] Regardless of whether the communication was a bullet or a penis, Cambridge Analytica demonstrates the same desire to penetrate that the older mass social engineers had, coupled with the more targeted focus of hacker social engineers.

And, like both the mass social engineers and professional social engineer penetration testers, Cambridge Analytica had to report results. After Trump's victory in the 2016 presidential election, the company's employees were highly sought after—if controversial—speakers at advertising and technology conferences. In 2017, Cambridge Analytica Head of Digital Molly Schweickert presented a summary of their Trump campaign strategy to d3con, and Nix presented at the Online Marketing Rockstars conference.[80] Both of these presentations were in Germany—a state with strong data protection laws—so the audience members seemed taken aback by the sheer glut of data Cambridge Analytica was working with, and they had critical questions for the two employees. However, both Schweickert and Nix were at the time basking in the glow of victory, so they fended off the criticisms and proudly reported on their penetration successes.

Masspersonal Social Engineering since 2016

We focused on the 2016 election efforts by the Internet Research Agency and Cambridge Analytica because they are well documented. As important as those cases are, in the face of the past six years, they may seem dated. Here, we point to more recent examples of potential masspersonal social engineering. We say "potential" because we do not yet have all the data on the post-2016 efforts, but we suspect all the elements of masspersonal social engineering (trashing, pretexting, bullshitting, and penetrating, with a fluid mix of interpersonal and mass communication) are present in them. With our heuristic in mind, let's consider some more recent, potential instances of masspersonal social engineering.

Russia

First, the Russians appear to have continued their masspersonal social engineering efforts after the 2016 election. A December 2018 report from the Office of the Director of National Intelligence detailed Russian "influence activities" and "messaging campaigns" meant to sway the 2018 mid-term election results.[81] The Russian operations followed much the same playbook detailed above, including using pretexts (a network of shell companies to hide their involvement, thousands of fake social media profiles) and efforts to target particular groups with messages meant to inflame political, racial, and cultural divisions. The difference in 2018, according to the DNI report, was that instead of trying to support one side or the other in the election, as they had in 2016 with their support for Donald Trump, the Russians took positions on all sides of the political spectrum in an attempt to sow general political chaos, confusion, and division.[82] While we do not know the full extent of the impact of these operations, we do know that "Russia appears to have engaged in more disruptions to democratic dialogue in 2018 than in 2016, not fewer" and that the threat was perceived as serious enough for the military's US Cyber Command to undertake operations to disrupt IRA operations targeting the 2018 election.[83]

The Russian effort continued into 2020. Less than two months before the 2020 US presidential election, the US Treasury Department sanctioned "four Russia-linked individuals for attempting to influence the US electoral process." Most prominent among them was Andrii Derkach, whom the US described as "a member of the Ukrainian Parliament" and "an active Russian agent for over a decade, maintaining close connections with the Russian Intelligence Services." The US alleged that Derkach had "waged a covert influence campaign centered on cultivating false and unsubstantiated narratives concerning US officials in the upcoming 2020 presidential election" and had "almost certainly targeted the US voting populace, prominent US persons, and members of the US government."[84]

The targets of Derkach were then-presidential candidate Joe Biden and his son, Hunter. The government claimed that Derkach and his associates engaged in bullshitting through "US media, US-based social media platforms, and influential US persons," spreading "misleading and unsubstantiated allegations that current and former US officials engaged in corruption, money laundering, and unlawful political influence in Ukraine." Derkach and his associates used pretexts in the form of "media front companies" such as Nabu Leaks, Era-Media TOV, Skeptic TOV, and Only News as part of their effort.[85] Derkach used these fronts, in part, to help in "publicizing leaked phone calls" meant to serve as evidence of the Bidens' alleged corruption.[86] Derkach was also later linked to the Hunter Biden laptop story that seemed meant to serve as the Trump campaign's "October surprise" during the 2020 election. One cannot help but notice the similarities to the 2016 "hack and leak" operation involving stolen Clinton campaign emails and WikiLeaks, only this time it was the so-called "hard drive from hell" that was the focus of American right-wing media.[87]

The overall impact of this most recent campaign is still difficult to assess. It seems safe to say that it did indeed have a significant impact. But that impact was likely the opposite of what the Russians had intended. The prominent persons and government officials targeted in the operation included President Trump's personal lawyer, Rudy Giuliani, as well as two GOP Senators, Chuck Grassley of Iowa and Ron Johnson of Wisconsin. Giuliani is reported to have met in December 2019 with Derkach and one of his associates in an effort to find incriminating evidence on the Bidens. Derkach's associate is reported to have sparked a Senate investigation led by Grassley and Johnson into the Bidens' dealings in Ukraine.[88] The US Treasury claimed in September 2020 that that investigation, along with a parallel investigation carried out in Ukraine, was "designed to culminate prior to election day" with the intended effect of undermining candidate Biden's election chances.[89] Of course, President Trump's push for Ukraine to investigate the Bidens was one of the grounds for his first impeachment in January 2020.[90] Neither the Ukraine investigation, nor the Senate investigation, uncovered any substantial evidence of wrongdoing on the part of the Bidens.[91] And, of course, the Hunter Biden laptop story fizzled almost immediately, gaining little traction in the mainstream press while major social media platforms for a time blocked users from sharing links to it.[92] In the end, getting the cooperation of the then-president's personal lawyer and an investigation out of the US Senate are significant impacts. Whether they ended up hurting Trump's 2020 prospects for reelection is unknown. But they certainly seem not to have helped: despite the Russian efforts, Trump lost the election.

Iran

Russia was not the only foreign state actor accused of attempting to interfere in the 2020 US election. In August 2020, the DNI also warned that China and Iran were attempting to use online influence campaigns to sow division and undermine confidence in the upcoming election.[93] Of the two, we know the most about the Iranian efforts because, quite frankly, they got caught. Twice. In both cases, the US government accused Iran of spreading bullshit messages, including "voter intimidation emails and dissemination of US election-related propaganda."[94]

In the first instance, beginning in mid-October, only a few weeks before the election, registered Democrats in Florida and Alaska began receiving emails from Iranian operatives pretexting as Trump-supporting "Proud Boys." The emails told these voters that the Proud Boys had gained access to voting infrastructure, knew that the individual was a Democrat, and threatened "we will come after you" if the recipient didn't change their party affiliation to Republican and vote for Trump.[95] The Iranians allegedly engaged in trashing, obtaining voter registration information from the public web, from private vendors, and through exploitation of vulnerabilities in election-related websites.[96] This sparked an FBI investigation that within a couple of days revealed that hackers working on behalf of Iran had been behind the emails, not the Proud Boys.[97]

In the second instance, the FBI accused Iran of being behind a post-election campaign of online threats and intimidation targeted at federal and state law enforcement, cybersecurity, and election officials, as well as executives of voting machine manufacturers. In this case, in mid-December a website titled, "Enemies of the People," appeared that identified a list of individuals who were accused of having helped swing the election results against then-President Trump. The site included pictures of the individuals with crosshairs superimposed over them, which was interpreted by some as a call for violence. This "hit list," as the *Washington Post* called it, "was shared on social media using the hashtags, #remembertheirfaces and #NoQuarterForTraitors."[98] We see pretexting in this case as well. In addition to the now-standard fake social media personas, the Iranian hackers are alleged to have "creat[ed] fictitious media sites and spoof[ed] legitimate media sites."[99] The Enemies of the People list was posted to several websites and promoted by at least ten social media accounts across multiple platforms. Technical analysis revealed that those who created the websites and posted the information "had gone to great lengths to mask their identity."[100]

Whether this campaign had any significant impact on the outcome of the election remains unclear as these two component operations were quickly identified and outed publicly. Whatever the impact, the Iranian case is significant because it allows us to observe the use of the masspersonal social engineering elements by a state actor other than Russia. It also demonstrates how tightly interconnected those elements can sometimes be. Each incident relied on a complex combination of trashing, penetrating, and pretexting to spread bullshit messages. Penetration of election-related websites enabled the trashing of voters, while penetration of a Proud Boys email account may have enabled the sending of pretextual voter intimidation messages.[101] The bullshit in these incidents was spread through a combination of mass and targeted means. The Enemies of the People list was a website accessible to anyone and amplified using multiple domains and social media accounts, while the pretextual Proud Boys emails were more individually targeted. Finally, the first incident indicates how difficult it can sometimes be to disentangle the elements of an operation and truly understand what is going on. While the US government initially thought a propaganda video depicting the hacking of election-related websites to obtain voter data was fake, officials later concluded that the hackers had indeed breached those sites.[102] Penetration can enable bullshitting. But sometimes, as the government's initial assessment should remind us, even a bullshit claim to have penetrated a system might be as damaging as a real breach.

Domestic Masspersonal Social Engineering

Cambridge Analytica may be no more, but its tactics live on in contemporary political communication firms and operations.

In 2020, based on data from Facebook, the Stanford Internet Observatory released an analysis of the Austin, Texas-based political consulting firm Rally Forge.[103] That year, Facebook had removed accounts associated with that firm due to their use of pretexts, including pages set up by fake persons. As the *Washington Post* reported, "Rally Forge ... was 'working on behalf' of Turning Point Action, an affiliate of Turning Point USA, the prominent conservative youth organization based in Phoenix, Facebook concluded."[104]

The Stanford Internet Observatory took the Facebook data and traced a web of accounts operating across Facebook, Instagram, and Twitter. Their analysis reveals that Rally Forge accounts used pretexts and bullshitting with the goal of penetrating in the form of building the appearance of conservative grassroots movements. As the report states, Rally Forge's pretexts created messaging "that appears to be driven by authentic grassroots energy is in fact sponsored by an undisclosed organization looking to sway public opinion."[105] The bullshitting on the part of Rally Forge included family-friendly messages, like "Take your kids hunting," alongside messages denying climate change, labeling Congressional Representative Alexandria Ocasio-Cortez an "invasive species," or mocking veganism. The report also indicates some success in penetration, with the "combined 36 Facebook Pages [having] 300,161 followers as of October 10, 2020."[106]

Conservative groups aren't the only ones engaging in potential masspersonal social engineering. Left-leaning groups have also been discovered using these tactics. Both the *New York Times* and the *Washington Post* have reported that several groups of progressive activists in Alabama used pretexts to undermine the candidacy of Roy Moore, a Republican candidate for the Senate.[107] The cyber security firm Yonder (formerly named New Knowledge) took what it learned when it analyzed the 2016 Russian efforts and applied it on a smaller scale in Alabama. Yonder created a Facebook page with a pretext: "they posed as conservative Alabamians, using it to try to divide Republicans and even to endorse a write-in candidate to draw votes from Mr. Moore."[108] Another activist, Matt Osborne, created a Facebook page based on another pretext. The page, "Dry Alabama," associated Moore's campaign with a campaign to ban alcohol in Alabama, hoping that such an association would frighten pro-business Republican voters.[109]

These efforts may have been coordinated by Project Birmingham, an overarching effort to apply Russian-style disinformation tactics in American politics.[110] Much as the older mass social engineers and more recent hacker social engineers argue, the justification by the activists using these tactics is that the Republicans are benefiting from masspersonal social engineering, so the Democrats may as well, too. As Osborne argued to the *Times*, "If you don't do it, you're fighting with one hand tied behind your back. You have a moral imperative to do this—to do whatever it takes."[111]

The reports about Rally Forge or Project Birmingham do not reveal much about the trashing phase of masspersonal social engineering. We can't confirm, but we can assume, that those operations gathered data on their targets. In the case of Phunware, a software company that was hired to develop the Trump/Pence 2020 mobile phone applications, we have a great deal of insight into masspersonal social engineering-style trashing practices.[112] Carrying on in the spirit of Cambridge Analytica, Phunware trashing practices were aggressive, wallowing in the glut of data available online and collecting over a billion smartphone device ID numbers.

As the *New Yorker* reported, these ID numbers "were collected from phones and tablets that use Phunware's software." However, Phunware also trashed via online advertising ventures, gathering data even in cases where it was not able to serve ads to internet users:

> According to people who have worked with the company, in addition to the data it obtains through its software, Phunware has been using its ad-placement business as a wholesale data-mining operation. . . . By collecting and storing this information, the company is able to compile a fairly comprehensive picture of every app downloaded on those devices, and any registration data a user has shared in order to use the app.[113]

Such trashing is geared towards the same microtargeting, penetrating vision that Cambridge Analytica held. Phunware's promotional materials boasted about its "1000+ mobile device segments that marketers can leverage to create custom audiences within the platform of their choice. Phunware segments are organized into six profile dimensions that enable marketers to quickly create audiences."[114]

Whether or not microtargeting works, it is certainly a selling point for firms like Phunware and the political campaigns that hire them.[115] There is every indication that not just microtargeting, but the broader suite of masspersonal social engineering tactics will continue to proliferate in the years ahead. Recent investigations by journalists, cybersecurity companies, and academics all point to a growing market for what is increasingly being referred to as "disinformation for hire." For example, a 2019 investigation by the Insikt Group, an arm of the threat intelligence company Recorded Future, reported on the availability and affordability of disinformation services on Russian underground forums.[116] But it's not just Russian cyber criminals in obscure corners of the internet willing to provide such services. There are an increasing number of public relations and marketing firms getting in on the action, too. In January 2021, "A BuzzFeed News review . . . found that since 2011, at least 27 online information operations have been partially or wholly attributed to PR or marketing firms. Of those, 19 occurred in 2019 alone."[117]

While that sounds bad, a report released that same month by the Computational Propaganda Project at the Oxford Internet Institute found that the problem might be even more widespread:

> Private firms increasingly provide manipulation campaigns. Over the last year, we found forty-eight instances of private companies deploying computational propaganda on behalf of a political actor. Since 2018 there have been more than 65 firms offering computational propaganda as a service. In total, we have found

almost US $60 million was spent on hiring these firms since 2009.[118]

It is a trend that they called "industrialized disinformation."

Conclusion

The social engineering conducted by the Russians or Cambridge Analytica was masspersonal. It took advantage of the masspersonal communicative affordances of corporate social media, moving from targeting an individual (Steve Bannon or John Podesta) to targeting crowds like voters in Texas or Black Americans. The next moment, these social engineers would engage with individuals again, this time on Facebook, Twitter, or Instagram, convincing them to arrange protests or meetings, and thus using them as proxies to build even more crowds. The communication alternated between mass messages meant for audiences of thousands and customized and personalized messages for a smaller pool of targets.

The tactics were straight out of the hacker or phone phreak playbook—trash, pretext, bullshit, penetrate—honed over decades of interpersonal con artistry. But the ambitions were, of course, redolent of the mass social engineers, aimed as they were at mastering crowds.

The Internet Research Agency and Cambridge Analytica activities remind us that, to a degree, we've been here before. As we saw in previous chapters, publicity, propaganda, public relations, and consent engineering all have deep ties to the national security state, whether in service to the state during wartime, as in the two world wars, or in efforts to safeguard democracy against incursions of nefarious outsiders during the Cold War. Cambridge Analytica's claims to precision are not so different from the dreams of precision targeting of undecided voters based on detailed personal data and "scientific polling" held at the start of the last century by Edward

M. House, Woodrow Wilson's long-time political advisor.[119] The IRA and Cambridge Analytica activities remind us too of decades of Russian and Soviet "active measures."[120] But they should also remind us of the longstanding intersections of what we now call "strategic communication" with wartime propaganda; the United States's own experience with political warfare during the Cold War, including interference in the domestic politics and elections of European allies; and more recent interest in the weaponization of social media for information warfare.[121] For example, in 2010, we got a small glimpse into the world of military contractor involvement in online propaganda and smear campaigns as a result of the hacker collective, Anonymous, exfiltrating and publicly releasing tens of thousands of emails from the defense contractor HBGary.[122] The following year, we learned about the US military's interest in developing technology for what they called "persona management," which would allow one operator to control potentially dozens of social media accounts, all with the aim of manipulating online discussion.[123] And as we have argued, Russian operations and Cambridge Analytica–like efforts are very likely still happening, with Iran and Phunware now operating similar campaigns. Indeed, things may well be worse now as the death toll from the COVID-19 pandemic rises and as we witness an attempted insurrection fomented by Trump and his supporters. The dream of swaying people, individually or collectively, to do one's bidding is an old one—and it's not going away anytime soon. To the degree that such techniques are portrayed, rightly or wrongly, as effective, others will likely continue to adopt and deploy them.

So, Cambridge Analytica and the IRA represent only a new phase of the old problems of social engineering, and their ilk appear to be gaining power. What is to be done about masspersonal social engineering? We turn to that question next.

8
Conclusion: Ameliorating Masspersonal Social Engineering

In this book, we took up the call from several people to consider emerging forms of manipulative communication as social engineering. In doing so, we argued that there is, indeed, a strong affinity between the concept of social engineering and the mixture of practices such as email hacks, meme production and sharing, attempts to microtarget voters, and social media pretexts. Not only are these practices the descendants of mass social engineering and propaganda developed in the early twentieth century, they are also related to the hacker practice of interpersonal con artistry.

We have deployed a critical approach to strategic communication through the use of concepts derived from hacker studies. This leads to a new understanding of our contemporary digital media environment. After tracing the threads between mass and hacker interpersonal social engineering, our penultimate ambition was to reveal elements of trashing, pretexting, bullshitting, and penetrating in two key events: the Russian election interference campaign and the use of psychographics by Cambridge Analytica to sway public political opinion. These were indications of what security researcher Thomas Rid calls "political engineering, social engineering on a strategic level."[1] These events—and many others like them that we still experience—saw hacker social engineering scaled up to societal level, and thus approaching the ambitions of the older mass social engineers to penetrate media markets or large-scale social institutions.

We hope that our work shows that the relationship between targeted, interpersonal social engineering and mass social engineering is fluid. This may help us overcome impasses in current analyses of manipulative communication practices, including those of firms like Cambridge Analytica. Do targeted, interpersonal techniques work? Or should we be more concerned about mass messaging? A 2020 debate held between two important scholars of political communication, Emma Briant and David Karpf, pitted these two perspectives against one another, with Briant taking up the argument that psychographic targeting of individuals is a dire threat to democratic deliberation, and Karpf expressing more concerns about mass messaging. To be fair to both of them, the debate was structured so that they had to simplify their positions by taking these opposing sides. However, if the debate was meant to resolve the impasse, it did not. Briant maintained that precise targeting is the most dangerous development in recent political communication, and Karpf argued that mass communication is still dominant in political communication.[2]

By bringing hacker concepts to bear on mass social engineering, and by bringing the results forward to our current media landscape, we may be able to avoid such mass/targeted impasses. After all, the mass social engineers saw themselves as crowdmasters—that is, interested in mass communication to convert unruly crowds into manageable publics. The hacker social engineers targeted individuals or, at most, organizations. If the two types of social engineering have many overlaps, then it is wise to speak of *masspersonal* social engineering, where social engineering can easily slide from large-scale campaigns to hyper-targeted individual hacks and back again, depending on the target to be penetrated, the interactions the social engineer seeks to have with the target, the specific goals the social engineer wants to achieve, and the specific platform or medium used for communication.[3] Our concept of masspersonal social engineering conforms to observations that our contemporary, digital media environment is a masspersonal communicative one.[4] The existence of a fluid form of masspersonal social engineering is supported by our analysis of the IRA and Cambridge Analytica practices, both of which oscillated between targeted and mass communication and relied on the techniques of trashing, pretexting, bullshitting, and penetrating.

But some questions remain. First of all, is masspersonal social engineering effective? For example, the burning question after the 2016 US election was, of course, did the Russians get Trump elected? Or was it Cambridge Analytica? Or something else? To be blunt, we don't think we can answer this question, but our analysis of masspersonal social engineering can provide clues, as we will discuss.

Second, can masspersonal social engineering be ethical? Throughout this book, we have discussed some disturbing examples of social engineering: blaming unionists for a fire that killed children; increasing smoking among women; covertly listening in on people's phone calls; breaking into computer networks; discouraging Black Americans from voting; using ill begotten data to craft highly targeted fear appeals; sowing chaos in the democratic process. Are there cases in which the coordinated use of trashing, pretexting, and bullshitting can be put to good penetrating use? To be blunt again, we don't think so.

Finally, what is to be done about masspersonal social engineering?

Is Masspersonal Social Engineering Effective?

There remains the question of whether masspersonal social engineering works. Our key cases have been the Russian election interference operation and the microtargeting political communication of Cambridge Analytica. Did either affect the US presidential election? Has the combination of big data, social media, and psychological sciences been able to deliver on the longstanding dream of precise, direct effects of communication?

Despite all the documentation of these events, academics and journalists are still trying to figure this out. For example, Kathleen Hall Jamieson's book, *Cyberwar*, emphatically argues that the Russian masspersonal social engineering campaign did work, setting the agenda for debate and swaying undecided voters in key swing states.[5] Similarly, the UK's *Channel 4* investigative report into Cambridge Analytica's use of data to help the Trump campaign argues for a correlation between their efforts to deter Black voters from voting and the reduced turnout among Black voters in key states like Wisconsin.[6] "We can't know what effect these ads had on each voter that saw them," the report notes, "but for the first time in 20 years, Black turnout fell."[7] Such efforts dovetail with the Russian pretexts of Black Lives Matter activists who repeatedly stated that voting was a useless gesture for Black Americans. The fact that Black voting was down in states like Wisconsin and Michigan lends some credence to the argument that such masspersonal social engineering worked.

However, there are strong arguments against this claim. For example, the Cambridge Analytica skeptic David Karpf argues that "Donald Trump's campaign didn't possess a secret data innovation. His unlikely victory was due to a messy confluence of factors," including a poor campaign by Clinton (she infamously did not campaign in Wisconsin during the general election), James Comey's announcement that the FBI was going to investigate Clinton's email server eleven days before the election, and old-fashioned sexism.[8] As for Russian interference, Harvard Law professor Yochai Benkler argues that "we should remain skeptical that Russians spending a couple of hundred thousand dollars on Facebook advertising had meaningful impact relative to a presidential campaign spending millions of dollars while being guided by Facebook's own marketing team."[9] Similarly, Thomas Rid, a leading scholar of cyber conflict, argued that it was "unlikely that the trolls convinced many, if any, American voters to change their minds" or "had any discernible effect on the voting behavior of American citizens."[10] Russia might have sought to help the Trump campaign, but Trump was also, in fact, campaigning, and it is too reductive to suggest the tens of millions who voted for Trump were duped by Russia.

The debate about the effectiveness of Cambridge Analytica or the IRA will no doubt continue to rage. We ourselves are not in a position to decide if either were successful. Our purpose is to conceptualize masspersonal social engineering. But in doing so, we find the need to sound three warnings.

First, over and over again in the information security literature, we find reports that *hacker* social engineering—the interpersonal trashing, pretexting, and bullshitting—is extremely effective in achieving the social engineer's goal of penetrating information systems.[11] This claim is not disputed. If highly targeted, interpersonal social engineering is effective, this suggests that messages customized for particular groups of targets—say, one of the thirty-two personality types theorized in the OCEAN psychographic system—may very well be effective, so long as the messages are based on extensive trashing, have pretexts that are recognizable to targets, and include carefully crafted bullshit. This would of course take incredible effort, but the success of interpersonal social engineering indicates that such a masspersonal campaign has the real potential to be effective.

We are aware that this warning reflects a longstanding assumption about the instrumental approach to communication: direct effects might not work just yet, but with advancements in technologies for data collection and targeting, one day we would be able to achieve such effects. From the days of mass social engineers Lee,

Bernays, and Fleischman to the present, a recurring dream has been about the advances in communication technologies and social science theories leading to a deeper and more accurate penetration of the minds of audiences. What we're suggesting here, however, is not that advances in technology will inevitably lead to the realization of the dream of direct effects. Instead, we're suggesting that future masspersonal social engineering may better implement the successful model of interpersonal hacker social engineering on a large scale. At that point, Cambridge Analytica or the IRA's efforts may seem to be clumsy approximations of those efforts.

Our second warning: our analysis of hacker social engineers shows that their success in gaining access to restricted systems stems from repeated efforts. For example, the hacker social engineer Kevin Mitnick's recollection of his social engineering attack on Motorola in the 1990s is marked by many, many phone calls to many employees—with a variety of pretexts—until he was able to find an employee who could give him access to the source code of a new cell phone.[12] This highly iterative approach is predicated on the fact that the social engineer's pretexts and bullshit won't work on everyone in a given organization, but that they will work on at least one person. That such repeated efforts have been scaled up in contemporary digital media should come as no surprise. Phishing emails operate on this principle. Likewise, digital advertising—which measures success in tiny increments—also reflects this principle. This alerts us to the fact that, in the case of masspersonal social engineering, if an entity is allowed to continue trashing, pretexting, and bullshitting a given target, it will find vulnerabilities in that target. And when the target in question is a large swath of society, there are many opportunities to attack it over and over again. If, for example, the target is a population that has some preexisting distrust in government institutions, and a masspersonal social engineer repeats messaging undermining faith in a particular governmental practice—say, conducting an election, addressing a pandemic, or preventing climate change—then the likelihood of success starts to rise. Russia's continuous efforts over the past several years are strikingly similar to Mitnick's repeated phone calls.

Finally, social engineering is not the only technique for shaping society. Again, to use hacker social engineering as a source for clues, hacker social engineers don't just rely on interpersonal con artistry. They can couple that with technical attacks—the most common connotation of "hacking"—such as exploiting software or network flaws. Likewise, organizations seeking to shape an election have more tools than just masspersonal social engineering at their disposal. In the case of Donald Trump and the Republican Party's work to suppress Black voting, voter ID laws, gerrymandering, and challenges to ballots from predominantly Black population centers have been effective techniques and can be used in concert with masspersonal social engineering efforts to suppress minority voting. Messages coming from Republican masspersonal social engineers are also amplified in what propaganda scholars call the "right-wing media ecosystem," including outlets such as Fox News, which has parroted the idea that votes coming from predominantly Black population centers should be suspected as fraudulent.[13] While we may or may not be able to tell if Cambridge Analytica's or Russia's efforts to suppress Black voting were successful, we can say without a doubt that they were done in concert with these other practices. And when it comes to the Russian effort, they are obviously not limited to masspersonal social engineering; Russian intelligence organizations have also hacked election systems.[14]

However, despite these warnings, we also accept the possibility that there is simply no way to effectively expand hacker social engineering to a mass scale, that there is some boundary between an interpersonal con job and the manipulation of a society. In fact, as we will discuss below, we ourselves subscribe to the vision of communication as the mutual constitution of reality—beset by struggles, to be sure, but fundamentally about the voices of people who seek to understand themselves and their worlds and, in doing so, associate together for the good of all. From this viewpoint, if efforts at masspersonal social engineering are not effective, another possible explanation could be that the fundamental assumptions about communication made by Russian operatives or political consultancies were always too simple and linear. Perhaps we still fail to achieve the dream of direct effects because, as critics have noted for decades, the role of communication in society and culture is not as simple as changing individual minds that then add up linearly and predictably into a mass. Perhaps the role of communication in society and culture is more complex and given to nonlinearity, feedback, dissensus, and emergence than the masspersonal social engineers would have us believe. It could be the case that an interpersonal hacker social engineer could fool a few of us for some of the time, but there could be no way for masspersonal social engineers to fool many of us, all of the time.

But we should be clear. None of this means that masspersonal social engineering or other forms of social engineering don't have effects. Perhaps they are not the effects that their users have so often dreamed of having. Just because they might not do what was intended, that doesn't mean they do nothing at all and that the effects they do have aren't potentially negative.

There are many examples of sociotechnical developments that did not have particularly great immediate effects or, in some cases, did not really work at all, but which are now recognized as having been vitally important despite (or perhaps because of) their failures. Since one of the penetration metaphors social engineers rely upon is that of the bullet, some military analogies are, perhaps, in order. We can think, for example, of historical analysis of the ironclad Civil War-era military ship the USS *Monitor* which argues that the ship was largely a failure as a warship but a resounding success in terms of the publicity it generated, as well as in its status as a powerful vision for the future of naval warfare.[15] Much later, during WWII, Norbert Wiener's "anti-aircraft predictor" failed as an effective anti-aircraft weapon but succeeded in sparking a "cybernetic worldview" that became the core of our understanding of the so-called Information Age.[16] In the postwar period, a preeminent historian of technology points to another "failed" weapon system for its important, long-term implications: the SAGE air defense system. Again, though the system did not really succeed at its primary mission of mitigating the Soviet nuclear threat, it was an important forerunner to the emergence of the internet.[17] Finally, we can look to Operation Igloo White during Vietnam. This was an effort to use electronic surveillance and precision airstrikes to stop infiltration along the Ho Chi Minh Trail. It didn't work. But the vision it inspired of intelligence-driven, precision targeting found a home in US military thinking from the 1980s on.[18] The details of these cases vary, but a pattern appears: the initial efforts to deploy the new weapon system largely failed, but in that failure, participants and observers alike were left with the question: But what if it had worked? What if we made some modifications and tried harder next time? Entire sociopolitical systems were built around these failures in order to achieve their intended effects.

We should expect nothing less with respect to masspersonal social engineering. Indeed, doubts about the effectiveness of the Russian effort to affect the 2016 US presidential election didn't stop the Russians from trying again in the mid-term elections of 2018 and the presidential election of 2020. Rather than fretting about whether or not Cambridge Analytica's targeting efforts worked, successors to that firm, like Phunware or Rally Forge, plowed ahead with their own efforts. These continued examples of attempted manipulation along the lines of what happened in 2016 indicate that there are plenty of people still striving to get masspersonal social engineering to work. Even if it fails, the effects of masspersonal social engineering can include renewed efforts at manipulative communication.

Moreover, if the stated purpose of these efforts was to sow distrust in institutions, the response to the COVID-19 pandemic and the insurrection attempt of January 6, 2021 indicate that the mission may well have succeeded. Therefore, we strongly suggest continued analysis of the efforts of masspersonal social engineers, whether they are effective or not.

Can Masspersonal Social Engineering Be Ethical?

A second question: can social engineering ever be used for good?

Throughout this book we have relied heavily on the insights of people who refer to themselves as "ethical social engineers." Sharon Conheady, Jenny Radcliffe, and Chris Hadnagy identify this way, offering their hacking services to organizations who want to test their security. Even the reformed felon hacker Kevin Mitnick offers his services as an ethical hacker.[19] These people most definitely engage in social engineering by our definition. They trash; they create pretexts; they bullshit; and they penetrate.

But, as we discussed in chapter 6, the fact that these social engineers report on their activities tells us something is different about ethical social engineering for penetration tests. There comes a point when the hacker drops the pretext, resumes their professionalized persona, and provides a report of their penetration test. Ideally, at that point, the organization can learn from the experience and offer training to employees. All of the elements of social engineering are in place, but only for a limited period of time, and the intention is always to reveal any deceptive practices to victims.

Something similar could be said about efforts to educate people about privacy and data manipulation. An example here is "Game Time!," an app made by security consultant Aelon Porat. "Game Time!" is a pretext, much in the same way that Cambridge Analytica's "Sex Compass" was. Under the guise of providing a diversion to the user, the "Game Time!" app gathered a tremendous amount of data about the user, from location to photographs to contact lists. However, unlike other apps, Porat's "Game Time!" also included a detailed and publicly visible log of the data collected. Porat uses that log to demonstrate to "Game Time!" players what those data could be used for: politically profiling the user, mapping their commute, and discerning their economic status based on their purchases.[20] All of this *comes from a game*. But, much like ethical social engineers, after Porat's app engages in trashing, pretexting, bullshitting, and penetrating, he drops the pretext to reveal the extent of the digital data his app collects.

So, in both the case of the ethical social engineers and privacy advocates like Porat, we may be manipulated, but very quickly they reveal their manipulations with the goal of opening our eyes to what social engineering can achieve.

And this is where the cases we have explored diverge. Masspersonal social engineering does not allow for pretexts to be dropped in time for targets to learn anything about manipulation. In the case of international conflicts, such as the Russian operation, there is, of course, never an intention to drop the pretext. Vladimir Putin repeatedly denies his government's involvement in the US election. At best, he claims again and again, private citizens were expressing their opinions online—nothing more.[21] At worst, Putin tells us, the masspersonal social engineering is an uncoordinated effort by individual "patriotic hackers."[22]

This is also true of Cambridge Analytica. To be fair, Cambridge Analytica employees did publicly present some of their tactics while they took a victory lap in 2017. People were justifiably curious as to how the company potentially helped elect a demagogue to office. However, even in the same presentations, they denied that the data they drew on were obtained unethically. When asked how Cambridge Analytica acquired their Facebook data, then-CEO Alexander Nix simply stated that any data that they had came from people "voluntarily giving up their data. They're doing this in full knowledge of what's happening. This is nothing Machiavellian or untoward."[23] Such denials came at an ever increasing pace as more revelations about the company hit the news. Indeed, we would likely not know the extent of Cambridge Analytica's social engineering if it weren't for researchers like Emma Briant, reporters like Carole Cadwalladr, and whistleblowers like Brittany Kaiser and Christopher Wylie.

And even if Cambridge Analytica or Russia revealed the full extent of their masspersonal social engineering, whatever damage they managed to achieve was already done. The votes were cast and four years of Trump ensued. Contrast this with the limited scope of an ethical social engineering engagement.

Although their consequences are on a smaller scale, similar concerns can be expressed about "stealth marketing" campaigns. Consider a campaign in 2010, when attractive actors were paid by Blackberry to sit in bars and flirt with men, with the goal of getting the men to handle the new Blackberry Pearl phone and punch their phone numbers into it.[24] Mixing market research, pretexts, and bullshit, the marketing practice is clearly masspersonal social engineering. Soon, of course, such stealth marketing practices shifted online, with influencers being paid to surreptitiously (in the parlance of marketing, "natively") shill products they purport to believe in.[25] Even if these stealth marketing firms reveal their manipulation of consumers at a later date—perhaps in some celebratory article in *Ad Age*—the damage is done. The dollars are spent. Intense but one-sided personal relationships were forged. And even if the masspersonal social engineering is done out of putative desire to learn how social engineering works—as the Democratic operatives claimed they were merely doing during a senate race in Alabama—the end result will not be new knowledge, but increasing distrust in institutions and a warrant for masspersonal social engineering to be normalized as a legitimate method of strategic communication.[26]

So, we may be able to accept social engineering for ethical purposes in the limited case of hacker social engineers conducting penetration tests, or in the case of demonstrations of the extent of privacy violations happening online, so long as they protect the privacy and reputations of anyone they interact with, operate with integrity, and, most importantly, drop their pretexts and report their results in public *before* any damage is done due to their social engineering. Professionals like Conheady, Radcliffe, Hadnagy, and Porat do these things. Without these revelations, however, we see no possibility of the ethical application of masspersonal social engineering.

It is true that social engineers hired to do penetration tests do so on behalf of powerful clients—the corporate managers who hire them. The information they glean about how well employees secure information can be abused by those employers, who may choose to fire employees who are manipulated by the social engineers.[27] Moreover, the overall solution that ethical social engineers offer—we hack you so you can learn not to be hacked—relies upon a logic of individual responsibility for security, rather than considering security as a more social, organizational issue that large institutions should take charge of. These are flaws in ethical social engineering, but we think that they can be addressed without doing away with the entire field of ethical hacker social engineering.

What to Do about Masspersonal Social Engineering?

If masspersonal social engineering is predominantly unethical, and if it has the potential to undermine democratic debate (that is, if it hasn't yet already), we need to find ways to ameliorate the damage masspersonal social engineering does.

Masspersonal social engineering is not merely the problem of disinformation. Rather, it is enabled by a collection of difficult, systemic challenges that span privacy and surveillance, cybersecurity, dark money in politics, the use and abuse of emerging ad tech, weak (or nonexistent) regulation of the influence industry, and more. Each of the elements of masspersonal social engineering relates directly to these more specific problems. Thus, we break down our suggestions based on each stage of masspersonal social engineering. Social engineering's reliance on these practices provides many avenues for mitigation. And even if masspersonal social engineering turns out to be ineffective, it's still worthwhile to consider solutions to problems like trashing, pretexting, bullshitting, and penetrating.

Trashing

The sheer lack of regulation of private data in the United States leads to gluts of information that enable trashing—or, to put it more politely, Big Data analysis—of unethically or illegally obtained personal information. Corporate social media sites are the main culprits here. But we cannot forget other, less well-known culprits: consumer profiling agencies like Acxiom, credit ratings firms, and governments, all of which collect as much data as possible under the assumption that, as communication scholar Mark Andrejevic puts it in his critique of unregulated data collection, "all data is potentially relevant no matter how seemingly trivial, irrelevant, personal, or invasive it may seem."[28]

It seems obvious to counsel people to avoid giving out their data—quit Facebook (and all of Facebook's subsidiaries) and Twitter, pay with cash, use a burner phone, use Signal, use Tor. Above all, we might point to frequent *Social-Engineer.org Podcast* guest Michael Bazzell's guides to becoming "digitally invisible."[29] As Andrejevic notes, "familiar norms of individual privacy threaten the data mining model to its core."[30] There are a host of personal privacy guides out there, and widespread adoption of their principles among individuals may in fact undermine trashing.

All of this seems fine until we think about how it puts an incredible burden on each of us to self-secure and take full responsibility for that security. Indeed, the advice to digitally disappear through the use of complex privacy tools and techniques like setting up shell companies in New Mexico echoes the advice given by professional hacker social engineers: it's up to *each of us* to secure *everything*. Be constantly aware of the information you share. In the extreme, trust no one.

Instead, we would argue for far better privacy laws and restrictions on how data are collected in the United States. Cambridge Analytica's Molly Schweickert implicitly acknowledges this solution in her post-mortem discussion of the 2016 Trump campaign. She presented her work in Germany, a country known for strong data protections, and her German audience was somewhat bewildered at the access to Americans' personal information Cambridge Analytica enjoyed. As she explained, the US relies on opting out of data collection. Every citizen must actively decline to have their personal data gathered. In contrast, Europe is opt-in.[31] This has not changed in the past several years—in fact, the 2020 Trump campaign was arguably even more invasive in its trashing practices than it was in 2016.[32] We suggest a key antidote to masspersonal social engineering can come in the form of making American companies adhere to far stricter data collection laws, with severe penalties for the sorts of abuses that Cambridge Analytica indulged in.

It is possible that the United States will catch up to the European Union in this regard and thus alleviate masspersonal social engineering. The March 2020 report from the US Congress's Cyberspace Solarium Commission explicitly connects the insecurity of personal data to the rise of social engineering. The report predicts that

> Authoritarian states will take advantage of preferred relationships with technology firms to build in backdoors for government access that allow them to surveil the private lives of citizens and political opponents at home and abroad. In addition to advertising,

propaganda will be micro-targeted and tailored to an individual based on personal data and search history.[33]

With such surveillance capacities in place,

> The information stolen from American entrepreneurs, public officials, industry leaders, everyday citizens, and even clandestine operatives is fueling social engineering and espionage campaigns against US firms and agencies.[34]

Due to increasing capacities for targeted propaganda and social engineering—that is, masspersonal social engineering—the report calls on the United States Congress to "pass a national data security and privacy protection law establishing and standardizing requirements for the collection, retention, and sharing of user data."[35] We concur, so long as the legislation includes directives to make the collection of personal information opt-in.[36]

One possible side effect of increased regulation of data-mining corporations may be less waste. As we have argued, trashing relies on a political economy of waste, where waste is a forgotten byproduct of consumerism. Likewise, surveillance capitalism relies in large part on our forgetting about our data. Just as the trash is picked up from the curb and taken . . . *somewhere* (do you know where your local dump is located?), so, too, do the data collection practices of corporate social media, app developers, and device software makers take our data *somewhere*. It's 2022. Do you know where your data are? Do you know where the nearest Google or Facebook data center is located? Perhaps you don't, but know this: it is massive, it consumes a great deal of energy and water, and it is not going to shrink or disappear so long as our personal data are easily obtained.[37]

Even if we severely curtail the future collection of personal data via legislation, there is still a great deal that has already been collected. We cannot forget this. To combat the data/trash/forgetting triad, we need a radically different relationship to the flows of information. Just as we need to grapple with the mounds of waste we are generating, we will have to grapple with the materiality of information as data are stored in massive server farms. As environmentalist and media studies scholar Mél Hogan argues in her criticism of data collection,

> While we equip ourselves with mass surveillance capabilities and are complicit in continuously generating data, we are not cognizant of the fact that our tracked bodies exist within a material world: one that is slowly compromised at the expense of being watched, detailed, and archived, in bits and numbers.[38]

In addition to an opt-in culture of personal data collection, there should be limits on how long the data are retained. And this means that the corporations who have gathered our data need to be required to delete our personal information rather than indefinitely store it—unless we affirmatively opt in to having a permanent record.

Pretexting

Role-playing, in itself, is not necessarily a problem. There are many instances in which people may want to play roles. A common example is the identity play that happens online. Such roles allow people to explore aspects of themselves that might in other contexts put them at risk. The cases where pretexts are used to harm other people—as in the case of interpersonal con artistry—are cases of fraud, plain and simple, and can be prosecuted as such. Otherwise, there is no pressing need to eliminate role-playing.

But the glut of "dark money" flooding politics—thanks in part to the US Supreme Court's 2010 *Citizens United v. FEC* decision—allows for well-funded pretexts on a national scale. Dark money "is money that has been routed through an opaque non-profit—thus concealing its true source from voters and investors alike."[39] This is, of course, a pretext straight out of the old mass social engineering playbook. Just as Edward Bernays, Doris Fleischman, and Ivy Lee might have created front groups to advocate positions on behalf of their clients, today's super PACs are advocating political positions for their wealthy and untraceable donors.[40] The explosion of dark money-funded groups can fundamentally shift politics away from the political parties and towards "'shadow parties'—organizations outside of the party that house the party elites."[41] These shadow parties will be beholden only to their unknown donors and can shape political programs by simply choosing which politicians get the money they need to campaign.

Compounding this is the eagerness for corporate social media to sell advertising space to anyone willing to pay for it. To rectify these problems, corporate social media sites may want to eradicate political advertising altogether if they cannot control who is buying the ads—particularly when the purchasers are coming from outside the state holding an election. But the temptation to get a slice of the now billions spent on electioneering is likely too great. Facebook argues that its "ad library" provides transparency in terms of political ad spending, but researchers have found that Facebook's attempt at self-regulation is woefully inadequate, rife with inconsistencies and missing information.[42] This allows for ads produced through pretexts to continue to flourish.

It's not just Facebook's self-regulation that is inadequate. Past mass social engineers have tried self-regulation and it simply does not work. The use of pretexts known as "front groups" has been discouraged by the Public Relations Society of America Volunteer Chapter since at least the 1950s because such pretexts create the false impression of popular support for positions held by powerful corporations and governments.[43] However, because public relations is a self-regulating industry, there are no effective ways to prevent public relations firms from setting up the same sorts of front groups Bernays and Fleischman were famous for. The use of front groups is only increasing as super PACs gain power and corporate social media seeks to sell access to their audiences. In the case of elections, since they in particular are fundamental public goods, then it follows that they should be publicly funded, full stop, with no private money being used to fund campaigns.[44]

Bullshitting

Perhaps the most difficult social engineering element to combat is bullshit. There is, simply put, a great deal of truth-indifferent communication that mixes deception, accuracy, and friendliness. From the moment we wake up and see our first advertisements on our phone until we fall asleep thinking about the day's political, economic, and cultural news, we will confront an overwhelming stream of bullshit. This is not to mention daily conversations with coworkers, friends, family, and strangers, all of which might contain elements of deception, accuracy, and friendliness.

The bullshit we encounter among friends, family, and co-workers may not have a true antidote, other than our occasionally calling it out. Recall Chandra Mukerji's study of the hitchhikers, which argues that bullshit is often about maintaining social bonds.[45] The social penalty for calling out our loved ones' bullshit might include estrangement from them. Holiday gatherings will probably be even more awkward.

Mediated bullshit, however, has an ancient adversary: media literacy. For every production of mediated bullshit, from advertising to public relations to contemporary disinformation campaigns, a host of scholars and activists leap up in challenge. The lessons provided by the Media Education Foundation in the 1990s are carried on by the work of the scholars at the Data & Society think tank and the Technology and Social Change Research Project at Harvard. The latter has given us the Media Manipulation Casebook, an excellent resource for decoding online misinformation campaigns.[46]

A notable example of anti-bullshit media literacy work is the #YourSlipIsShowing hashtag movement of Black feminists. Unlike the more formal projects, #YourSlipIsShowing is more of an ad-hoc effort, but one that exposed many of the same sorts of pretexts the

Russians and others would go on to use. In a remarkable feat of internet research, these Black feminists traced fake Twitter accounts, which purported to be the accounts of Black activists, to their 4chan progenitors, who turned out to be racists, men's rights activists, and pickup artists. However, despite the important work of #YourSlipIsShowing, this movement has largely gone unsung: "one of the earliest crowdsourced anti-misinformation campaigns on the internet has been mostly ignored by the mainstream media."[47]

In addition to dissecting mediated bullshit as the #YourSlipIsShowing activists did, we must also question the systems in which it thrives. We have to ask critical questions as to why this bullshit is before us and why it displaces other, non-bullshit messages.[48] As media literacy scholars have argued since at least the 1990s,

> The goals of a loosely regulated, commercial media have no educational, cultural, or informational imperatives. As much of the literature on the political economy of the media suggests, they are there to maximize profits and to serve a set of corporate interests.[49]

We should consider media as systems, not as loose collections of texts. This perspective orients us to what we might call the political economy of bullshit.

Knowing the economics of bullshit takes on a new urgency in these days of masspersonal digital media, where text is increasingly customized for each individual and where conspiracy theories and misinformation reign supreme—and where the sources of such messages might be obfuscated. As one contemporary guide to media literacy argues, "it is necessary to rethink media literacy in the age of platforms."[50] While we undoubtedly agree, rethinking media literacy needs to go further. Social media platforms are a big part of the problem, but as we have argued in this book, masspersonal social engineering implicates even greater systemic problems at the intersections of technology and politics. Media literacy in the age of masspersonal social engineering requires literacy about the problems that allow for trashing, pretexting, and penetrating, too.

Media literacy to combat masspersonal social engineering is not just about media but also about privacy, surveillance, cybersecurity, and more. And, it also requires us to take back ownership of social media, perhaps in the form of community-owned, non-profit social media systems.[51]

This brings us back to the sort of bullshit we might get from friends, family, and co-workers. This sociable bullshit takes on newer, potentially damaging forms when it is channeled through corporate social media. It shifts from relatively harmless, interpersonal discussions to masspersonal media—especially because Facebook and other corporate social media are built to amplify messages our contacts share with us.[52] Our relatives become vehicles for bullshit and hence possible vectors for masspersonal social engineering. Tackling mediated bullshit—especially as it appears on corporate social media, as contemporary media literacy advocates urge us to do—may help us address local, familial bullshit.

Penetrating

Of course, the most obvious penetration-related problem we face is the woeful state of cybersecurity among public and private institutions in the United States. While US cybersecurity discourse for years focused on the potential for catastrophic cyber doom scenarios leading to physical destruction and loss of life, the events of 2016 and later—including recent massive hacks of corporate and government networks as part of the 2021 SolarWinds incident—have driven home the message that the dominant cyber threats are informational. Cybersecurity is threatened by espionage, intellectual property theft, and the over-collection and misuse of personal data.[53] That is, it is threatened by trashing. But in addition to the trashing-related suggestions mentioned above, we can mitigate such threats by, in essence, doing a better job of locking the doors, preventing systems that hold such data from being penetrated in the first place. Once again, we suggest taking seriously the many proposals found in the recent Cyberspace Solarium Commission Report in this regard.[54]

But the penetration metaphor has many fathers, including schools of communication that have produced generations of people who conceive of communication as a game of penetration. As the former Cambridge Analytica employee turned whistleblower Christopher Wylie put it, from this perspective,

> culture change can be thought of as nudging the distribution curve of culture up or down. What the data allowed us to do was to disaggregate that culture into individuals, who became movable units of that society.[55]

This view of culture emerging as the sum of communicative nudges of individuals is precisely what historian of communication Christopher Simpson called the "science of coercion," or "communication as domination."[56] More polite terms include "strategic communication" or "media effects," where carefully calibrated messages can master crowds or shape individuals' opinions. Such thinking leads directly to the epistemology of masspersonal social engineering, which would elevate the penetrating effects of communication over all other considerations. Although he's talking about seducing women by any means possible, the pickup artist Ross Jeffries summarizes this view quite well: "the purpose of your communication is not to give her an understanding. The purpose of your communication is to get you a result!"[57] Add up all these "results," this line of thinking goes, and a social order emerges.

Moreover, this communication-as-penetration model slips easily into the mouths of people who would characterize all communication as a military endeavor. While this book is deeply indebted to security researchers, we find that too often the language they reach for is militarized and securitized. Just as we have offered an explanation of the process of masspersonal social engineering, there have been several other attempts in recent years to describe the stages of information operations, fake news, disinformation, social media manipulation, and malignant foreign influence campaigns.[58] For example, security analyst Bruce Schneier takes up defense contractor Lockheed Martin's concept of the "cyber kill chain" to theorize the cycle of information operations. In drawing on researchers like Schneir and offering our own process, we appear to be offering a "cybersecurity kill chain" of our own. However, we would reject this characterization. This kind of language is a result of a narrow vision of how to talk about what we're calling social engineering. The conceptual universe of "cyber" has been so thoroughly militarized that, as soon as something is framed as cybersecurity, everything about it—from diagnosis to solution—becomes a military problem. The penetration metaphor of communication—where messages are bullets—has become the dominant language for talking about all things cyber, including disinformation.[59]

This approach has, of course, been criticized for generations. Communication theorist James Carey famously called it the "transmission model" of communication.[60] Christopher Simpson's "science of coercion" is a harsher condemnation. They join cultural studies scholar Raymond Williams, who noted in the 1960s that

> it is indeed monstrous that human advances in psychology, sociology, and communication should be used or thought of as powerful techniques *against* people, just as it is rotten to try to reduce the faculty of human choice to 'sales resistance'.... Much of this talk of weapons and impact is the jejune bravado of deeply confused men [sic].[61]

Our contribution to this critique is modest: we simply want to join the ranks of all those who maintain communication is richer and more than the penetration metaphors of penises and bullets would have us believe. Against these penetrating metaphors, we turn to colleagues who repeatedly envision a better model of communication.

Instead of communication as penetration, we agree with communication scholar Guobin Yang's argument for "communication as translation." Drawing on insights from a range of thinkers, including Williams, Carey, Walter Benjamin, and Patricia Hill Collins, Yang argues communication-as-translation "is premised on the recognition of difference, dialogue, receptivity, mutual change, and self-transformation."[62] Like translation across languages, communication as translation is a series of utterances that are bounded in meaning (they relate back to the original language or statement), but at the same time, they are boundless in interpretation. We are constantly translating one another—even if we speak the same language. Moreover, as Yang argues, if we acknowledge communication as an act of translation, we ought to be open and listen to people who do not share our backgrounds, most especially people whose voices are not often translated to the mainstream.

Yang's approach echoes Black feminist scholar Patricia Hill Collins's call for dialogue and coalition-building across autonomous groups.[63] Collins uses the concept of a multiplicity of stories to elaborate on this approach, noting how storytellers are "writing one immense story, with different parts of the story coming from a multitude of different perspectives."[64] The gathering up of such narratives could "form constellations illuminating experiential particularity," a "democratic communication that is inclusive without suppressing particularity."[65] In this vision, the "universal become[s] a fluid and emergent terrain of agreement, a space for becoming in difference."[66] This is opposed to a predetermined universal being imposed upon the multiplicity of people's stories—the eradication of difference, a Big Translation of all ideas into one dominant story.

As tempting as it might sound, this vision of emergent, from-below storytelling, dialogue, and coalition-building is decidedly not data-driven. Particular stories are not data points to be aggregated into a Big Dataset that can allow for analysis without theory. Rather, the narratives that are being produced are coming from autonomous groups—including and especially marginalized groups—who are defining themselves and then actively articulating their views with other groups. We echo Collins's warning that autonomous groups, such as Black feminists, must define themselves and not be defined by others:

> Because self-definition is key to individual and group empowerment, ceding the power of self-definition to other groups, no matter how well-meaning or supportive of Black women they may be, in essence replicates existing power hierarchies. . . . As Audre Lorde points out, "it is axiomatic that if we do not define ourselves for ourselves, we will be defined by others—for their use and our detriment."[67]

Moreover, narratives are not things to be controlled, penetrated, or dominated. Narratives are collective, locally negotiated and defined, and then translated into new contexts, bearing with them traces of their previous meanings while being open to new meanings.

Yang's communication-as-translation, then, is not about dialogue across differences theorized by those who—including the adherents of the communication-as-penetration model—would sort people into specific types and use such typologies to craft messages to manipulate them. Rather, it is about the "complex unity in difference" that emerges as such autonomous groups build coalitions through dialogue.[68] Such complex unity has space for the social engineers, perhaps, but it does not endorse their limited vision of communication as penetration.

To put this another way, we do not share the faith of those who want to use "data for good." Even the critics of Cambridge Analytica, the whistleblowers Brittany Kaiser and Christopher Wylie, hold onto the view that enough data in the hands of good elites can lead to social and economic justice.[69] Kaiser's own eagerness to search for "data for good" led her to praise Phunware in the pages of her book, calling it "a Big Data company that is returning the data they hold to consumers and rewarding them for its use."[70] Phunware would go on to make the Trump 2020 campaign app, by some measures one of the most invasive apps ever used in US politics.[71] This desire to use communication as penetration for good echoes the views of the mass social engineers, who arose during the early twentieth century and held that scientific "facts" (presented in the right way) could be used to implement a perfected society. The parallels between the social reformers of the early twentieth century and progressive data scientists of the early twenty-first century are startling: they all start with an earnest belief that data can inform social justice, but they gravitate towards people in power in the blind belief that at best those powerful actors will support their data- or fact-driven social justice goals, or at worst they can take the money they earn from their penetration campaigns and someday use it for good. No matter their desire to do good—through health campaigns, or public service campaigns, etc.—the penetrative communication approach is readily appropriated by those in power.

When those in power inevitably use data and communication to maintain and strengthen social hierarchies and inequalities, we act surprised. We should not be. And this leads us to a critique of those who take a more critical approach to communication, as we do. If the communication-as-domination view places too much emphasis on effects, critical scholars have often placed too little emphasis on thinking more expansively about the effects of communication. Clearly, communication, whether mass, interpersonal, or masspersonal, has effects. But we need to think more about effects as broader than just the linear penetration of individual minds and the changing of individual behavior (e.g., political opinion and voting) to consider communication effects as nonlinear and emergent phenomena that exist and are important, but that may not be amenable to the quantitative social science methodologies that predominate in traditional media effects research. Thus, we support recent efforts by communication scholars to take a more critical approach to the study of strategic communication, as well as to create a "critical media effects framework."[72]

Again, however, let us repeat that each of these practices—trashing, pretexting, bullshitting, and penetrating—are not in themselves masspersonal social engineering. Their concatenation is the problem. We are wary of anyone who trashes us. We are wary of anyone who uses a pretext. We are most definitely wary of people who bullshit us. And we are wary of people who seek to penetrate us. But we should be most concerned about the masspersonal social engineers who put all of these practices together into a new form of manipulative communication.

Fortunately, any chain can be broken at any given link. What we offer here is, admittedly, only a start to how we might do that. We invite others to help further our understanding and options for breaking the chain of malicious use of masspersonal social engineering.

Notes

Introduction

1. Hannah Miao, "Democratic Senators Formally Request Investigation of Hawley and Cruz after Deadly Capitol Insurrection," CNBC, January 21, 2021, https://www.cnbc.com/2021/01/21/capitol-riot-democrats-file-ethics-complaint-against-cruz-hawley.html.

2. Kelsey Vlamis, "A Texas Man Who Tweeted 'Assassinate AOC' Is Facing Charges in the Capitol Riot, FBI Says," *Business Insider*, January 23, 2021, https://www.businessinsider.com/texas-man-who-tweeted-assassinate-aoc-charged-in-capitol-riot-2021-1; Jay Reeves, Lisa Mascaro, Calvin Woodward, Dustin Weaver, and Michael Casey, "Capitol Riot More Sinister than It Looked as Gallows, Pipes and Guns Turn Up," *Tampa Bay Times*, January 11, 2021, https://www.tampabay.com/news/florida-politics/2021/01/11/capitol-riot-more-sinister-that-it-looked-as-gallows-pipes-and-guns-turn-up/.

3. "Assessing Russian Activities and Intentions in Recent US Elections," Washington, DC: Office of the Director of National Intelligence, 2017, https://www.dni.gov/files/documents/ICA_2017_01.pdf.

4. Renee DiResta, Kris Shaffer, Becky Ruppel, David Sullivan, Robert Matney, Ryan Fox, Jonathan Albright, and Ben Johnson, "The Tactics & Tropes of the Internet Research Agency," New Knowledge [now Yonder], 2018; Philip N. Howard, Bharath Ganesh, Dimitra Liotsiou, John Kelly, and Camille François, "The IRA, Social Media and Political Polarization in the United States, 2012–2018," Oxford: Computational Propaganda Research Project, 2018.

5. Erin Gallagher, "Introduction: Memetic Warfare," *Medium*, July 29, 2018, https://medium.com/@erin_gallagher/alt-right-culture-jamming-and-memetic-warfare-93b646263f7d.

6. Nina Jankowicz, "How an Anti-Trump Flash Mob Found Itself in the Middle of Russian Meddling," POLITICO, July 5, 2020, https://www.politico.com/news/magazine/2020/07/05/how-an-anti-trump-flash-mob-found-itself-in-the-middle-of-russian-meddling-348729; DiResta et al., "The Tactics & Tropes of the Internet Research Agency," 21.

7. Issie Lapowsky, "25 Geniuses Who Are Creating the Future of Business," *Wired*, April 26, 2016, https://www.wired.com/2016/04/wired-nextlist-2016/.

8. For example, Kathleen Hall Jamieson, *Cyberwar: How Russian Hackers and Trolls Helped Elect a President: What We Don't, Can't, and Do Know* (New York: Oxford University Press, 2018); Yochai Benkler, Robert Faris, and Hal Roberts, *Network Propaganda: Manipulation, Disinformation, and Radicalization in American Politics* (New York: Oxford University Press, 2018).

9. "Foreign Threats to the 2020 US Federal Elections," Washington, DC: National Intelligence Council, March 10, 2021, https://www.dni.gov/files/ODNI/documents/assessments/ICA-declass-16MAR21.pdf; Lara Seligman, "Mattis Confirms Russia Interfered in U.S. Midterm Elections," *Foreign Policy*, December 1, 2018, https://foreignpolicy.com/2018/12/01/mattis-confirms-russia-interfered-in-us-midterm-elections-putin-trump/; Adam Goldman, Julian E. Barnes, Maggie Haberman, and Nicholas Fandos, "Lawmakers Are Warned That Russia Is Meddling to Re-Elect Trump," *New York Times*, February 20, 2020, https://www.nytimes.com/2020/02/20/us/politics/russian-interference-trump-democrats.html.

10. Sam Schechner, Emily Glazer, and Patience Haggin, "Political Campaigns Know Where You've Been. They're Tracking Your Phone," *Wall Street Journal*, October 10, 2019, https://www.wsj.com/articles/political-campaigns-track-cellphones-to-identify-and-target-individual-voters-11570718889; Stanford Internet Observatory, *Reply-Guys Go Hunting: An Investigation into a U.S. Astroturfing Operation on Facebook, Twitter, and Instagram*, Palo Alto, CA: Stanford University, 2020, https://cyber.fsi.stanford.edu/io/news/oct-2020-fb-rally-forge; Scott Shane and Alan Blinder, "Secret Experiment in Alabama Senate Race Imitated Russian Tactics," *New York Times*, December 20, 2018, https://www.nytimes.com/2018/12/19/us/alabama-senate-roy-jones-russia.html.

11. Insikt Group, "The Price of Influence: Disinformation in the Private Sector," Recorded Future, September 30, 2019, https://www.recordedfuture.com/disinformation-service-campaigns/.

12. Sean T. Lawson, *Cybersecurity Discourse in the United States: Cyber-Doom Rhetoric and Beyond*. (London: Routledge, 2020), 116.

13. Philip N. Howard, *Lie Machines: How to Save Democracy from Troll Armies, Deceitful Robots, Junk News Operations, and Political Operatives* (New Haven, CT: Yale University Press, 2020).

14. Peter Pomerantsev, *This Is Not Propaganda: Adventures in the War against Reality* (New York: PublicAffairs, 2019).

15. Samuel Woolley and Philip N Howard, *Computational Propaganda: Political Parties, Politicians, and Political Manipulation on Social Media* (Oxford: Oxford University Press, 2019).

16. Cory Wimberly, *How Propaganda Became Public Relations: Foucault and the Corporate Government of the Public* (New York: Routledge, 2020).

17. Benkler, Faris, and Roberts, *Network Propaganda*.

18. "Assessing Russian Activities and Intentions in Recent US Elections."

19. David M. Beskow and Kathleen M. Carley, "Social Cybersecurity: An Emerging National Security Requirement," *Military Review: The Professional Journal of the U.S. Army*, April 2019, 118.

20. Tate Ryan-Mosley, "The Technology That Powers the 2020 Campaigns, Explained," *MIT Technology Review*, September 28, 2020.

21. Jamieson, *Cyberwar*.

22. Brandon Valeriano and Ryan C. Maness, *Cyber War versus Cyber Realities: Cyber Conflict in the International System* (Oxford: Oxford University Press, 2015); Michael N. Schmitt, ed., *Tallinn Manual 2.0 on the International Law Applicable to Cyber Operations*, 2nd ed. (Cambridge: Cambridge University Press, 2017).

23. "Treasury Takes Further Action Against Russian-Linked Actors," US Department of the Treasury, accessed January 20, 2021, https://home.treasury.gov/news/press-releases/sm1232; Donie O'Sullivan and Alex Marquardt, "Iranian Hackers Who Posed as the Proud Boys Accessed Voter Data in One State, Feds Say," CNN.com, October 31, 2020, https://www.cnn.com/2020/10/30/politics/iran-hackers-proud-boys/index.html; Craig Silverman, Jane Lytvynenko, and William Kung, "Disinformation For Hire: How a New Breed of PR Firms Is Selling Lies Online," *BuzzFeed News*, January 6, 2020, https://www.buzzfeednews.com/article/craigsilverman/disinformation-for-hire-black-pr-firms.

24. Garance Burke, "Financially Troubled Startup Helped Power Trump Campaign," *AP News*, November 17, 2020, https://apnews.com/article/phunware-app-helped-power-trump-campaign-89ed273f60e37ff9ee020dd2f5d3df04.

25. Amy Goodman, Emma Briant, Karim Amer, Brittany Kaiser, and Jehane Noujaim, "The Weaponization of Data: Cambridge Analytica, Information Warfare

& the 2016 Election of Trump," *Democracy Now!*, January 10, 2020, video, 59:02, https://www.democracynow.org/2020/1/10/defense_contractors_are_using_a_new.

26. Sean Lawson, "Beyond Cyber-Doom: Assessing the Limits of Hypothetical Scenarios in the Framing of Cyber-Threats," *Journal of Information Technology & Politics* 10, no. 1 (January 1, 2013): 86–103, https://doi.org/10.1080/19331681.2012.759059; Sean Lawson, "Putting the 'War' in Cyberwar: Metaphor, Analogy, and Cybersecurity Discourse in the United States," *First Monday* 17, no. 7 (2012), https://firstmonday.org/article/view/3848/3270.

27. Sharon Conheady, *Social Engineering in IT Security: Tools, Tactics, and Techniques* (New York: McGraw-Hill Education, 2014), 1.

28. Raphael Satter, Jeff Donn, and Chad Day, "Inside Story: How Russians Hacked the Democrats' Emails," AP News, November 4, 2017, https://apnews.com/dea73efc01594839957c3c9a6c962b8a.

29. Dean Takahashi, "Kevin Mitnick: An Interview on Trump, Russians, and Blockchain with the World's Most Famous Hacker (Updated)," *VentureBeat* (blog), July 28, 2018, https://venturebeat.com/2018/07/28/kevin-mitnick-an-interview-on-trump-russians-and-blockchain-with-the-worlds-most-famous-hacker/.

30. Matthew Rosenberg, Nicole Perlroth, and David E. Sanger, "'Chaos Is the Point': Russian Hackers and Trolls Grow Stealthier in 2020," *New York Times*, January 10, 2020, https://www.nytimes.com/2020/01/10/us/politics/russia-hacking-disinformation-election.html.

31. Thomas Rid, quoted in Max Fisher, "Russian Hackers Find Ready Bullhorns in the Media," *New York Times*, January 8, 2017, https://www.nytimes.com/2017/01/08/world/europe/russian-hackers-find-ready-bullhorns-in-the-media.html; Also see Megan Dunham Keim, "Your Facts Are Not Safe with Us: Russian Information Operations As Social Engineering," Hacking Illustrated Series InfoSec Tutorial Videos, presented at BSides, Philadelphia, 2017, video, 50:28, https://www.irongeek.com/i.php?page=videos/bsidesphilly2017/bsidesphilly-cs01-your-facts-are-not-safe-with-us-russian-information-operations-as-social-engineering-meagan-dunham-keim.

32. Clint Watts, *Messing with the Enemy: Surviving in a Social Media World of Hackers, Terrorists, Russians, and Fake News* (New York: HarperCollins, 2018).

33. Davide Andreoletti and Enrico Frumento, "Social Engineering to the Extreme: The Cambridge Analytica Case," Dogana Project, March 23, 2018, https://www.dogana-project.eu/index.php/social-engineering-blog/11-social-engineering/92-cambridge-analytica.

34. Christopher Wylie, *Mindf*ck: Cambridge Analytica and the Plot to Break America* (New York: Random House, 2019), 82, 92.

35. Justin Sherman and Anastasios Arampatzis, "Social Engineering as a Threat to Societies: The Cambridge Analytica Case," RealClearDefense, July 18, 2018, https://www.realcleardefense.com/articles/2018/07/18/social_engineering_as_a_threat_to_societies_the_cambridge_analytica_case_113620.html.

36. Wimberly. *How Propaganda Became Public Relations*, 2.

37. "Assessing Russian Activities and Intentions in Recent US Elections," 3.

38. Harold D. Lasswell, "The Theory of Political Propaganda," *American Political Science Review* 21, no. 3 (August 1927): 627, https://doi.org/10.2307/1945515.

39. "Assessing Russian Activities and Intentions in Recent US Elections."

40. "Assessing Russian Activities and Intentions in Recent US Elections," 9.

41. Edward L. Bernays, "The Engineering of Consent" (Philadelphia, 1947).

42. Brittany Kaiser, *Targeted: The Cambridge Analytica Whistleblower's Inside Story of How Big Data, Trump, and Facebook Broke Democracy and How It Can Happen Again* (New York: HarperCollins, 2019), 12.

43. Kaiser, 14.

44. "Assessing Russian Activities and Intentions in Recent US Elections," 3.

45. DiResta et al., "The Tactics & Tropes of the Internet Research Agency," 13.

46. Kaiser, *Targeted*, 21.

47. Carole Cadwalladr and Emma Graham-Harrison, "How Cambridge Analytica Turned Facebook 'Likes' into a Lucrative Political Tool," *Guardian*, March 17, 2018, https://www.theguardian.com/technology/2018/mar/17/facebook-cambridge-analytica-kogan-data-algorithm.

48. Bill Allison and Misyrlena Egkolfopoulou, "Trump Outpaces Biden in Zeroing In on Voters with Facebook Tools," Bloomberg.com, July 13, 2020, https://www.bloomberg.com/news/articles/2020-07-13/trump-more-than-biden-is-tapping-into-facebook-targeting-tools.

49. Patrick B. O'Sullivan, "Bridging the Mass-Interpersonal Divide Synthesis Scholarship in HCR," *Human Communication Research* 25, no. 4 (1999): 569–588, https://doi.org/10.1111/j.1468-2958.1999.tb00462.x.

50. Alvin Toffler, *Future Shock (New York: Bantam, 1970)*; Alvin Toffler and Heidi Toffler, *Revolutionary Wealth* (New York: Knopf, 2006).

51. B. J. Fogg, "Mass Interpersonal Persuasion: An Early View of a New Phenomenon," *Persuasive Technology*, 2008, 23–34.

52. Patrick B. O'Sullivan and Caleb T. Carr, "Masspersonal Communication: A Model Bridging the Mass-Interpersonal Divide," *New Media & Society* 20, no. 3 (March 1, 2018): 1161–1180. https://doi.org/10.1177/1461444816686104.

53. Megan French and Natalya N. Bazarova, "Is Anybody out There?: Understanding Masspersonal Communication through Expectations for Response across Social Media Platforms," *Journal of Computer-Mediated Communication* 22, no. 6 (2017): 303–319.

54. O'Sullivan and Carr, "Masspersonal Communication," 1166.

55. O'Sullivan and Carr, 1168.

56. French and Bazarova, "Is Anybody out There?"

57. O'Sullivan and Carr, "Masspersonal Communication," 1172.

58. Whitney Phillips, *This Is Why We Can't Have Nice Things: Mapping the Relationship between Online Trolling and Mainstream Culture* (Cambridge, MA: MIT Press, 2015).

59. Ryan-Mosley, "The Technology That Powers the 2020 Campaigns, Explained."

60. Joseph M. Hatfield, "Social Engineering in Cybersecurity: The Evolution of a Concept," *Computers & Security* 73 (March 1, 2018): 102, https://doi.org/10.1016/j.cose.2017.10.008.

61. In particular, see John M. Jordan, *Machine-Age Ideology: Social Engineering and American Liberalism, 1911–1939*, new ed. (Chapel Hill, NC: The University of North Carolina Press, 2010); and Stuart Ewen, *PR! A Social History of Spin* (New York: Basic Books, 1996).

62. Gerald Stanley Lee, *Crowds: A Moving-Picture of Democracy* (Garden City, NY: Doubleday Page, 1913), http://www.gutenberg.org/ebooks/15759.

Chapter 1

1. Gerald Stanley Lee, *Crowds: A Moving-Picture of Democracy* (Garden City, NY: Doubleday Page, 1913), 57, http://www.gutenberg.org/ebooks/15759.

2. Stuart Ewen, *PR! A Social History of Spin* (New York: Basic Books, 1996), 9.

3. Ewen, 10.

4. Ewen, 11.

5. Samuel C. Florman, *The Existential Pleasures of Engineering* (New York: St. Martin's, 1996).

6. Florman, 3.

7. Florman, 6.

8. Stanley K. Schultz and Clay McShane, "To Engineer the Metropolis: Sewers, Sanitation, and City Planning in Late-Nineteenth-Century America," *The Journal of American History* 65, no. 2 (1978): 389–411, https://doi.org/10.2307/1894086.

9. John M. Jordan, *Machine-Age Ideology: Social Engineering and American Liberalism, 1911–1939*, new ed. (Chapel Hill, NC: The University of North Carolina Press, 2010), 53.

10. John F. McClymer, *War and Welfare: Social Engineering in America, 1890–1925* (Westport, CT.: Greenwood Press, 1980); Jordan, *Machine-Age Ideology*; Ewen, *PR! A Social History of Spin*.

11. Gustave Le Bon, *The Crowd: A Study of the Popular Mind* (London: Ernest Benn, 1896); Walter Lippmann, *Drift and Mastery: An Attempt to Diagnose the Current Unrest* (New York: Mitchell Kennerley, 1914); Lee, *Crowds*.

12. Le Bon, *The Crowd*, xvii.

13. Joseph S. Davis, "Statistics and Social Engineering," *Journal of the American Statistical Association* 32, no. 197 (1937): 1–7.

14. Edwin Lee Earp, *The Social Engineer* (New York, Eaton & Mains; Cincinnati, Jennings & Graham, 1911), 27, http://archive.org/details/cu31924014043370.

15. Earp, xi.

16. Earp, xv, our emphasis.

17. McClymer, *War and Welfare*, 12.

18. McClymer, 14.

19. Paul Underwood Kellogg, *The Pittsburgh Survey: Findings in Six Volumes* (New York: Charities Publication Committee, 1910); *Social Engineering in Cincinnati: The Annual Report of the Council of Social Agencies of Cincinnati* (Cincinnati, OH: The Council of Social Agencies of Cincinnati, 1919).

20. McClymer, *War and Welfare*, 219.

21. McClymer, 12.

22. McClymer, 39.

23. McClymer, 46.

24. McClymer, 43.

25. Henry S. Dennison, *Organization engineering* (New York: McGraw-Hill, 1931), 3.

26. Jordan, *Machine-Age Ideology*, 35.

27. Harry Braverman, *Labor and Monopoly Capital; the Degradation of Work in the Twentieth Century* (New York: Monthly Review Press, 1974).

28. Jordan, *Machine-Age Ideology*, 37.

29. Morris L. Cooke, "The Spirit and Social Significance of Scientific Management," *Journal of Political Economy* 21, no. 6 (June 1913): 483, https://doi.org/10.1086/252258.

30. Braverman, *Labor and Monopoly Capital*, 72; However, as Dennis Tourish notes, it is very likely that Taylor made his data about "Schmidt" up, basing him loosely on Hungarian workers he was observing. Dennis Tourish, *Management Studies in Crisis: Fraud, Deception and Meaningless Research* (Cambridge: Cambridge University Press, 2019), 14–15.

31. *Original Films of Frank B. Gilbreth*, 1945, video, 32:13, http://archive.org/details/0809_Original_Films_of_Frank_B_Gilbreth_02_12_34_00.

32. Quoted in Jordan, *Machine-Age Ideology*, 54.

33. Mario Krenn, "From Scientific Management to Homemaking: Lillian M. Gilbreth's Contributions to the Development of Management Thought," *Management & Organizational History* 6, no. 2 (May 1, 2011), https://doi.org/10.1177/1744935910397035.

34. Jill Lepore, "Not So Fast," *The New Yorker*, October 5, 2009, https://www.newyorker.com/magazine/2009/10/12/not-so-fast.

35. A key site for the fusion of social reform and scientific management was Chicago and its suburbs, particularly Cicero, Illinois. See Kyle Bruce, "Henry S. Dennison, Elton Mayo, and Human Relations Historiography," *Management & Organizational History* 1, no. 2 (May 1, 2006): 177–199, https://doi.org/10.1177/1744935906064095; John S. Hassard, "Rethinking the Hawthorne Studies: The Western Electric Research in Its Social, Political and Historical Context," *Human Relations* 65, no. 11 (October 1, 2012): 1431–1461, https://doi.org/10.1177/0018726712452168; Tourish, *Management Studies in Crisis*.

36. McClymer, *War and Welfare*, 46.

37. Nancy K. Bristow, *Making Men Moral: Social Engineering during the Great War* (New York: New York University Press, 1996).

38. Lepore, "Not So Fast."

39. Ewen, *PR! A Social History of Spin*, 67.

40. Scott M. Cutlip, *The Unseen Power: Public Relations, A History* (Hillsdale, NJ: Lawrence Erlbaum, 1994), 24.

41. Attributed to the press agent Dexter Fellows, quoted in Cutlip, 42.

42. Ewen, *PR! A Social History of Spin*, 76.

43. Edward L. Bernays and Howard Walden Cutler, "Theory and Practice of Public Relations: A Resume," in *The Engineering of Consent* (Norman: University of Oklahoma Press, 1955), 9.

44. Ivy Ledbetter Lee, *Human Nature and the Railroads* (E. S. Nash, 1915), 54, http://archive.org/details/humannatureandr01leegoog.

45. Lee, 22.

46. For extensive documentation on the influence of crowd psychology theorists on Lee and Bernays, see Wimberly, *How Propaganda Became Public Relations*.

47. Ewen, *PR! A Social History of Spin*, 76; Lee, *Crowds*.

48. Lee, 526.

49. Earp appears to have a stronger sense of broad education—delivered through churches and Sunday schools—leading to social engineering. This contrasts with Gerald Lee's more elitist view, a view that his cousin and others would embrace. See Earp, *The Social Engineer*.

50. Ewen, *PR! A Social History of Spin*, 81.

51. Lee, *Human Nature and the Railroads*, 20.

52. Edward L. Bernays and Mark Crispin Miller, *Propaganda*, first paperback ed. (Brooklyn: Ig Publishing, 2004), 39, 47; Edward L. Bernays, "The Engineering of Consent," *The ANNALS of the American Academy of Political and Social Science* 250, no. 1 (March 1947): 113–120, https://doi.org/10.1177/000271624725000116; Doris E. Fleischman, "Public Relations and the Consumer," presented at the Fashion Group, New York, October 30, 1935.

53. For discussions of how Doris Fleischman influenced Edward Bernays's ideas and contributed to their public relations efforts, see Margot Opdycke Lamme, "Outside the Prickly Nest: Revisiting Doris Fleischman," *American Journalism* 24, no. 3 (July 1, 2007): 85–107, https://doi.org/10.1080/08821127.2007.10678080; Susan Henry, "'There Is Nothing in This Profession . . . That a Woman Cannot Do': Doris E. Fleischman and the Beginnings of Public Relations," *American Journalism* 16, no. 2 (April 1, 1999): 85–111, https://doi.org/10.1080/08821127.1999.10739176.

54. Bernays and Miller, *Propaganda*, 71.

55. Bernays, "The Engineering of Consent," 114, 118.

56. Bernays and Miller, *Propaganda*, 76–78.

57. Bernays, "The Engineering of Consent," 113, 115.

58. Bernays and Miller, *Propaganda*, 37.

59. Bernays, "The Engineering of Consent," 114.

60. Bernays, 116.

61. Edward L. Bernays, *Crystallizing Public Opinion (1923)*; Walter Lippmann, *Public Opinion* (New York: Harcourt, Brace, 1922); For an analysis of how Bernays took up the ideas of Lippmann—and how Lippmann did not endorse Bernays's application—see Sue Curry Jansen, "Semantic Tyranny: How Edward L. Bernays Stole Walter Lippmann's Mojo and Got Away With It and Why It Still Matters," *International Journal of Communication* 7 (April 30, 2013). As Jansen argues, "Lippmann was consistently critical of the manipulation of public opinion by wartime propaganda and the transfer of propaganda techniques to peacetime endeavors. Conversely, Bernays contends that propaganda has positive social value in creating unified purpose in wartime and agreement on industrial purposes in peacetime" (1103).

62. Doris E. Fleischman, *An Outline of Careers for Women; a Practical Guide to Achievement* (Garden City, NY: Doubleday, Doran, 1928); Also see Lamme, "Outside the Prickly Nest," 94.

63. Fleischman, "Public Relations and the Consumer," 4.

64. Fleischman, 6.

65. Omar Ahmad Abu Arqoub, Bahire Efe Özad, and Adeola Abdulateef Elega, "The Engineering of Consent: A State-of-the-Art Review," *Public Relations Review* 45, no. 5 (December 1, 2019), https://doi.org/10.1016/j.pubrev.2019.101830.

66. Bernays, "The Engineering of Consent," 119.

67. See Mark Crispin Miller's introduction to Bernays and Miller, *Propaganda*, 21.

68. Bernays and Miller, 21.

69. Gene Marine, *America the Raped: The Engineering Mentality and the Devastation of a Continent* (New York: Simon & Schuster, 1969).

70. Florman, *The Existential Pleasures of Engineering*, 18.

71. Cutlip, *The Unseen Power*.

72. William F. Buckley Jr., "Our Mission Statement," *National Review*, November 19, 1955, https://www.nationalreview.com/1955/11/our-mission-statement-william-f-buckley-jr/.

73. Kevin Phillips, *Mediacracy: American Parties and Politics in the Communications Age* (Garden City, NY: Doubleday, 1975).

74. Phillips, 35.

75. James Taranto, "The Right's Happy Warrior," *Wall Street Journal*, April 30, 2010.

76. C. Wright Mills, *The Sociological Imagination* (New York: Oxford University Press, 1959), 113.

77. Mills, 113.

78. Like many historians of public relations, Ewen unfortunately overlooks the role of Doris Fleischman. Fleischman, like many of the women we will discuss in this book including Susy "Thunder" Headley and to a lesser extent Sharon Conheady), is largely overlooked in relation to male practitioners.

79. Stuart Ewen, *Captains of Consciousness: Advertising and the Social Roots of the Consumer Culture* (New York: McGraw-Hill, 1976).

80. Ewen, 53.

81. Ewen, 187.

82. Ewen's 1970s-era critique of consent engineering would return in a more forceful manner in his 1996 book, *PR! A Social History of Spin*.

83. W. Howard Chase, "Nothing Just Happens, Somebody Makes It Happen," *Public Relations Journal* (November 1962): 5.

84. Kirk Hallahan, Derina Holtzhausen, Betteke van Ruler, Dejan Verčič, and Krishnamurthy Sriramesh, "Defining Strategic Communication," *International Journal of Strategic Communication* 1, no. 1 (March 22, 2007): 6.

85. Zeynep Tufekci, "Engineering the Public: Big Data, Surveillance and Computational Politics," *First Monday* 19, no. 7 (July 2, 2014), http://www.firstmonday.dk/ojs/index.php/fm/article/view/4901.

Chapter 2

1. SN, Chesire Catalyst, and Emmanuel Goldstein. "Hackers on Planet Earth: Social Engineering." *2600: The Hacker Quarterly*, 1994. Video, 1:00:02. http://archive.org/details/HOPE-1-Social_Engineering.

2. Joseph M. Hatfield, "Social Engineering in Cybersecurity: The Evolution of a Concept," *Computers & Security* 73 (March 1, 2018): 102–113, https://doi.org/10.1016/j.cose.2017.10.008.

3. Tom Wolfe, "The 'Me' Decade and the Third Great Awakening," *New York Magazine*, August 23, 1976, https://nymag.com/news/features/45938/.

4. Fred Turner, *From Counterculture to Cyberculture: Stewart Brand, the Whole Earth Network, and the Rise of Digital Utopianism*, Chicago: University of Chicago Press, 2008.

5. Turner, 72.

6. Turner, 74.

7. Eva Illouz, *Saving the Modern Soul: Therapy, Emotions, and the Culture of Self-Help* (Berkeley: University of California Press, 2008), 155.

8. Richard Bandler, John Grinder, and Steve Andreas, *Frogs into Princes: Neuro Linguistic Programming* (Moab, UT: Real People Press, 1979); Judith DeLozier and John Grinder, *Turtles All the Way Down: Prerequisites to Personal Genius* (n.p.: Metamorphous Press, 1996); Richard Bandler, John Grinder, Steve Andreas, and Connirae Andreas, *Reframing: Neuro-Linguistic Programming [Trade Mark Symbol] and the Transformation of Meaning* (Moab, UT: Real People Press, 1982).

9. Joseph M. Reagle Jr., *Hacking Life: Systematized Living and Its Discontents* (Cambridge, MA: MIT Press, 2019).

10. This is similar to another, perhaps more well-known individualized therapy system, Werner Erhard's "est." See Sheridan Fenwick, *Getting It: The Psychology of Est* (Philadelphia: Lippincott, 1976), http://archive.org/details/gettingitpsychol00fenw.

11. Bandler, Grinder, and Andreas, *Frogs into Princes*.

12. Dana L. Cloud, *Control and Consolation in American Culture and Politics: Rhetorics of Therapy* (Thousand Oaks, CA: Sage Publications, 1998).

13. Cloud; Mark Satin, *New Age Politics: The Emerging New Alternative to Liberalism and Marxism* (Gearhart, OR: Fairweather Press, 1976).

14. Satin, *New Age Politics*, 2.

15. David Harvey, *A Brief History of Neoliberalism* (Oxford: Oxford University Press, 2005).

16. Adam Smith, *The Wealth of Nations* (New York: Bantam Classics, 2003).

17. Milton Friedman and Rose D. Friedman, *Capitalism and Freedom* (Chicago: University of Chicago Press, 2002); Gordon Tullock, *The Selected Works of Gordon Tullock*, ed. Charles Rowley (Indianapolis: Liberty Fund, 2006); Geoffrey Brennan and James M. Buchanan, *The Reason of Rules: Constitutional Political Economy*, Vol. 10 of *The Collected Works of James M. Buchanan* (Indianapolis: Liberty Fund, 2000); Samuel Edward Konkin III, "New Libertarian Manifesto" (Huntington Beach, CA: Koman Publishing, 1983), http://agorism.info/docs/NewLibertarianManifesto.pdf.

18. The earliest printed reference to this form of "social engineering" we could find is from 1984 in a computer bulletin board post by BIOC Agent 003: BIOC Agent 003, "Course in Basic Telecommunications Part IV," April 13, 1984, https://textfiles.com/phreak/BIOCAGENT/bioc2.wri. Another early mention is in *2600* magazine, later that year: Curious, "Dear 2600," *2600: The Hacker Quarterly*, June 14, 1984. Both of these sources were intended for a phreaking and hacking audience. Notably,

neither source defines the term, but uses it in passing, although the term is in quotes. This indicates that the term was a novel yet familiar one among phreaks and hackers.

Other sources (Phil Lapsley, *Exploding the Phone: The Untold Story of the Teenagers and Outlaws Who Hacked Ma Bell* (New York: Grove, 2014); Hatfield, "Social Engineering in Cybersecurity"; Ralph Lee, dir., *Secret History of Hacking*, produced by Mira King. September Films, 2001, video, 50:08, uploaded to YouTube 2013, https://www.youtube.com/watch?v=PUf1d-GuK0Q) draw on interviews with John "Cap'n Crunch" Draper, who argues he and Denny Teresi were using the term in the mid-1970s. Given that its earliest appearance in print was used as if the readers understood its meaning, this date of origin seems quite plausible. The phone phreaks enjoyed setting up telephone conferences amongst themselves, and they also held conventions in the 1970s. These were places where the term "social engineering" was likely used.

Also note that the phreaks and hackers used the term "bullshitting" more often than variations on "social engineering" to discuss their practice of conning Bell operators. We will discuss that further in the "Bullshitting" chapter.

19. Maureen Orth, "For Whom Ma Bell Tolls Not," *Los Angeles Times*, October 31, 1971, 29.

20. John Monberg and Stuart L. Esrock, "What a Long, Strange Trip Its Been: The Past, Present, and Uncertain Future of Universal Service," *Convergence: The International Journal of Research into New Media Technologies* 6, no. 4 (December 1, 2000): 81, https://doi.org/10.1177/135485650000600406.

21. Lapsley, *Exploding the Phone*.

22. "Testimony of Susan Headley, Tujunga, Calif.," in *Computer Security in the Federal Government and the Private Sector: Hearings before the Subcommittee on Oversight of Government Management of the Committee on Governmental Affairs, United States Senate, Ninety-Eighth Congress, First Session, October 25 and 26, 1983*, 24, Washington, DC: Government Printing Office, 1983, http://hdl.handle.net/2027/pst.000012047208.

23. For a collection of recordings of 1960s and '70s calls to various parts of the world in order to hear the phone equipment, visit http://www.wideweb.com/phonetrips/.

24. Lapsley, *Exploding the Phone*.

25. Henry M. Kluepfel, "Foiling the Wiley Hacker: More than Analysis and Containment," in *Proceedings. International Carnahan Conference on Security Technology*, 15–21, 1989, https://doi.org/10.1109/CCST.1989.751947; E. Gabriella Coleman and Alex Golub, "Hacker Practice: Moral Genres and the Cultural Articulation of

Liberalism," *Anthropological Theory* 8, no. 3 (2008): 255–277; Douglas Thomas, *Hacker Culture* (Minneapolis: University of Minnesota Press, 2002).

26. "The Electronic Delinquents," ABC, April 22, 1982.

27. Jim Russell, "Sorry, The Telephone Company You're Dailing Has Been Temporarily Disconnected," 16, Washington, DC: National Public Radio, January 1973, http://explodingthephone.com/docs/dbx0659.pdf.

28. Lee, *The Secret History of Hacking*.

29. The Mentor, "Engineering," *Phoenix Project BBS*, July 13, 1988, http://www.textfiles.com/messages/phoenix1.msg.

30. "Gin, n.1," in *OED Online*. Oxford University Press, accessed January 7, 2019, http://www.oed.com/view/Entry/78357.

31. Edmund Spenser, "The Faerie Queene."

32. Jessica Wolfe, "Nature and Technê in Spenser's Faerie Queene," in *A Companion to Tudor Literature*, ed. Kent Cartwright, (Hoboken: Blackwell, 2010), 412–427.

33. Verizon, *Data Breach Investigations Report*, 2018; Verizon, *Data Breach Investigations Report*, 2019.

34. Kevin D. Mitnick, *The Art of Deception: Controlling the Human Element of Security* (Hoboken: Wiley, 2002).

35. Claude S. Fischer, *America Calling: A Social History of the Telephone to 1940* (Berkeley: University of California Press, 1992), 66.

36. Fischer, 77.

37. A. B. Clark, "The Development of Telephony in the United States," *Transactions of the American Institute of Electrical Engineers, Part I: Communication and Electronics* 71, no. 5 (November 1952): 348, https://doi.org/10.1109/TCE.1952.6371872.

38. Nancy K. Baym, Yan Bing Zhang, and Mei-Chen Lin, "Social Interactions Across Media: Interpersonal Communication on the Internet, Telephone and Face-to-Face," *New Media & Society* 6, no. 3 (June 1, 2004): 299–318, https://doi.org/10.1177/1461444804041438; Charles R. Berger, "Interpersonal Communication," in *The International Encyclopedia of Communication*. International Communication Association, 2010, https://doi.org/10.1002/9781405186407.wbieci077.

39. Hatfield, "Social Engineering in Cybersecurity."

40. Jon Alexander and K. H. W. Schmidt, "Social Engineering: Genealogy of a Concept," in *Social Engineering*, eds. Adam Podgórecki, Jon Alexander, and Rob Shields, (Ottawa: Carleton University Press, 1996), 1.

41. Christopher Hadnagy, *Social Engineering: The Science of Human Hacking*, 2nd ed. (Indianapolis: Wiley, 2018), 7.

42. Stuart Ewen, *PR! A Social History of Spin* (New York: Basic Books, 1996), 9.

43. See the 6.58 mark of SN, Chesire Catalyst, and Emmanuel Goldstein, "Hackers on Planet Earth: Social Engineering," *2600: The Hacker Quarterly*, 1994, video, 1:00:02. http://archive.org/details/HOPE-1-Social_Engineering.

44. Hatfield, "Social Engineering in Cybersecurity."

45. Sharon Conheady, *Social Engineering in IT Security: Tools, Tactics, and Techniques* (New York: McGraw-Hill Education, 2014); Also see Hadnagy, *Social Engineering*, 2011.

46. Patrick B. O'Sullivan and Caleb T. Carr, "Masspersonal Communication: A Model Bridging the Mass-Interpersonal Divide," *New Media & Society* 20, no. 3 (March 1, 2018): 1161–1180, https://doi.org/10.1177/1461444816686104.

47. O'Sullivan and Carr, 1162.

Chapter 3

1. Quoted in Edward Humes, *Garbology: Our Dirty Love Affair with Trash* (New York: Penguin, 2013), 128.

2. John Hoffman, *The Art & Science of Dumpster Diving* (Port Townsend, WA: Loompanics, 1993), 130.

3. Michelle Slatalla and Joshua Quittner, *Masters of Deception: The Gang That Ruled Cyberspace* (New York: HarperCollins, 1995), 6, http://archive.org/details/mastersofdecepti00slat.

4. Slatalla and Quittner, 7.

5. John Reynolds, "Trashing the Phone Company," *Telephone Electronics Line 2*, no. 5 (May 1975), http://pdf.textfiles.com/zines/TEL/tel-07.pdf; A. Ben Dump, "Computing for the Masses: A Devious Approach," *TAP*, February 1980; Spenser Michaels, "The Philadelphia Story, Part 1," *TAP*, August 1982, http://www.textfiles.com/tap/issue.76/tap7602.jpg; Fred Steinbeck, "Gibberish," *TAP*, November 1982, http://www.textfiles.com/tap/issue.79/tap7903.jpg; The Kid & Co. and The Shadow, "More on Trashing: What to Look For, How to Act, Where to Go," *2600: The Hacker Quarterly* 1, no. 9 (September 1984), http://textfiles.com/phreak/TRASHING/trash.phk; "Some Thoughts on 'Garbage Picking,'" *2600: The Hacker Quarterly*, February 1984; Grifter, "Dumpster Diving: One Man's Trash . . ." Archive, Textfiles.com, 2002, http://web.textfiles.com/hacking/dumpster_diving.txt.

6. "The Electronic Delinquents," 20/20. ABC, April 22, 1982, 7.

7. "Testimony of Susan Headley, Tujunga, Calif.," in *Computer Security in the Federal Government and the Private Sector: Hearings before the Subcommittee on Oversight of Government Management of the Committee on Governmental Affairs, United States Senate, Ninety-Eighth Congress, First Session, October 25 and 26, 1983*, 24, Washington, DC: Government Printing Office, 1983, http://hdl.handle.net/2027/pst.000012047208.

8. Lars Eighner, "On Dumpster Diving," *The Threepenny Review*, no. 47 (1991): 6–8.

9. Slatalla and Quittner, *Masters of Deception*, 22.

10. Alfred R. Root and Alfred C. Welch, "The Continuing Consumer Study: A Basic Method for the Engineering of Advertising," *Journal of Marketing* 7, no. 1 (July 1942): 3–21, https://doi.org/10.2307/1246447.

11. Martin V. Melosi, *Garbage in the Cities: Refuse, Reform, and the Environment*, revised ed., History of the Urban Environment (Pittsburgh: University of Pittsburgh Press, 2005), 70–73.

12. Humes, *Garbology*, 57.

13. Melosi, *Garbage in the Cities*, 76.

14. Vance Packard, *The Waste Makers* (Philadelphia: David McKay, 1960), 3.

15. Packard, 10.

16. Humes, *Garbology*, 60.

17. Packard, *The Waste Makers*, 58.

18. Vance Packard, *The Hidden Persuaders* (London: Longmans, Green & Company, 1957).

19. "Bernays was paid by Dixie Cups to promote the sales of disposable plastic cups, and he did so by linking the imagery of an overflowing cup with subliminal images of vaginas and venereal diseases . . ." Alan Bilton, *Silent Film Comedy and American Culture* (New York: Springer, 2013), 16. However, we should sound a note of caution here. Early twentieth-century Dixie Cups were paper, not plastic. Moreover, Bilton does not cite a source here. That said, Dixie Cups were specifically marketed as more sanitary than "shared" cups, like those at a soda fountain, and the imagery of their advertising in the 1920s through '30s does suggest death and disease (if not subliminal images of vaginas). See "III: 'Beware the Common Drinking Cup'—Reform and the Assault on the Common Drinking Cup," Disposable America (blog), December 8, 2015, https://disposableamerica.org/course-projects/a-wholesome-drink/section-iii-beware-the-common-drinking-cup-progressive-reform-and-the-assault-on-the-common-drinking-cup/.

20. "VI: 'No Soda Fountain Needs to Be a "Germ Exchange"'—Soda Fountains, Restaurants, and Fast Food," Disposable America (blog), December 8, 2015, https://

disposableamerica.org/course-projects/a-wholesome-drink/section-vi-no-soda-fountain-needs-to-be-a-germ-exchange-soda-fountains-restaurants-and-fast-food/.

21. Humes, *Garbology*, 62.

22. Tim Jackson, "Live Better by Consuming Less?: Is There a 'Double Dividend' in Sustainable Consumption?" *Journal of Industrial Ecology* 9, no. 1–2 (2008): 19–36, https://doi.org/10.1162/1088198054084734.

23. Lisa Gitelman, *Paper Knowledge: Toward a Media History of Documents* (Durham, NC: Duke University Press, 2014), 5.

24. Mary Douglas, *Purity and Danger: An Analysis of Concepts of Pollution and Taboo* (London: Penguin, 1966).

25. John Scanlan, *On Garbage* (London: Reaktion, 2005), 8.

26. Scanlan, 62.

27. Scanlan, 41.

28. Myra J. Hird, "Knowing Waste: Towards an Inhuman Epistemology," *Social Epistemology* 26, no. 3–4 (2012): 465, https://doi.org/10.1080/02691728.2012.727195.

29. Humes, *Garbology*; Dietmar Offenhuber, *Waste Is Information: Infrastructure Legibility and Governance* (Cambridge, MA: MIT Press, 2017).

30. Offenhuber, 3.

31. Hird, "Knowing Waste," 455.

32. Colin Koopman, *How We Became Our Data: A Genealogy of the Informational Person* (Chicago: University of Chicago Press, 2019).

33. The Kid & Co. and The Shadow, "More on Trashing."

34. Steinbeck, "Gibberish."

35. Michaels, "The Philadelphia Story, Part 1."

36. "The Electronic Delinquents."

37. Hoffman, *The Art & Science of Dumpster Diving*, 128.

38. Christopher Hadnagy, *Social Engineering: The Art of Human Hacking* (Indianapolis: Wiley, 2010).

39. Richard Maxwell and Toby Miller, *Greening the Media* (New York: Oxford University Press, 2012); Mél Hogan, "Data Flows and Water Woes: The Utah Data Center," *Big Data & Society* 2, no. 2 (July 2015), https://doi.org/10.1177/2053951715592429.

40. David A. Umphress, "Diving the Digital Dumpster: The Impact of the Internet on Collecting Open-Source Intelligence," *Air & Space Power Journal* 19, no. 4 (2005): 82–91; Jason Fitterer, "Putting a Lid on Online Dumpster-Diving: Why the Fair and Accurate Credit Transactions Act Should Be Amended to Include E-Mail Receipts," *Northwestern Journal of Technology and Intellectual Property* 9, no. 8 (2011): 591–606; Michael Z. Green, "Against Employer Dumpster-Diving for Email," *South Carolina Law Review* 64, no. 2 (2012): 323–368; Nanna Bonde Thylstrup, "Data Out of Place: Toxic Traces and the Politics of Recycling," *Big Data & Society* 6, no. 2 (July 1, 2019), https://doi.org/10.1177/2053951719875479.

41. Mél Hogan, "The Archive as Dumpster," *Pivot: A Journal of Interdisciplinary Studies & Thought* 4, no. 1 (2015): 30, https://doi.org/10.25071/2369-7326.39565.

42. Packard, *The Waste Makers*.

43. Hird, "Knowing Waste"; Elizabeth V. Spelman, "Combing Through the Trash: Philosophy Goes Rummaging," *The Massachusetts Review* 52, no. 2 (2011): 313–325.

44. Thylstrup, "Data Out of Place."

45. Tim Harford, "Big Data: A Big Mistake?" *Significance* 11, no. 5 (2014): 14–19, https://doi.org/10.1111/j.1740-9713.2014.00778.x.

46. Sauvik Das and Adam Kramer, "Self-Censorship on Facebook," *Proceedings of the Seventh International AAAI Conference on Weblogs and Social Media* (2013): 8.

47. Zack Whittaker and Natasha Lomas, "Even Years Later, Twitter Doesn't Delete Your Direct Messages," *TechCrunch* (blog), February 15, 2019, http://social.techcrunch.com/2019/02/15/twitter-direct-messages/.

48. "Find What's Changed in a File—Computer—Docs Editors Help," Google, 2020, https://support.google.com/docs/answer/190843?co=GENIE.Platform%3DDesktop&hl=en.

49. Spelman, "Combing Through the Trash," 320.

50. "Testimony of Susan Headley, Tujunga, Calif.," 24.

51. "Some Thoughts on 'Garbage Picking,'" 4.

52. Matthew Kirschenbaum's analysis of materiality of information and digital forensics demonstrates the difficulty of truly deleting data. Data can be recovered using Magnetic Force Microscopy directly from hard drives—even ones that have been smashed or burned. See Matthew G. Kirschenbaum, *Mechanisms: New Media and the Forensic Imagination* (Cambridge, MA: MIT Press, 2012), chap. 1.

53. Johnny Long, *No Tech Hacking: A Guide to Social Engineering, Dumpster Diving, and Shoulder Surfing* ed. Scott Pinzon (Rockland, MA: Syngress, 2008), 121.

54. anonymous, "'Index of' inurl:'/$Recycle.Bin/'" Exploit Database, May 5, 2017, https://www.exploit-db.com/ghdb/4463/; anonymous, "'index of' inurl:recycler," Exploit Database, May 4, 2004, https://www.exploit-db.com/ghdb/205/; anonymous, "inurl:trash intitle:index.of," Exploit Database, June 6, 2016, https://www.exploit-db.com/ghdb/4295/.

55. Michael Bazzell, "142-OSINT Extravaganza and Book Release!" 2019, in The Privacy, Security, and OSINT Show, MP3 audio, 30:30, https://soundcloud.com/user-98066669/142-osint-extravaganza-and-book-release.

56. James Temperton, "Ashley Madison Data Posted Online—and It's Worse than Anyone Thought," Wired UK, August 19, 2015, http://www.wired.co.uk/news/archive/2015-08/19/ashley-madison-hack-data-leaked-online.

57. Martin Bos, David Kennedy, Justin Elze, and Scott White, "Ashley Madison Hacked. Dump Released," *TrustedSec* (blog), August 19, 2015, https://www.trustedsec.com/blog/ashley-madison-database-dumped/.

58. Dell Cameron, "It's Scary How Much Personal Data People Leave on Used Laptops and Phones, Researcher Finds," Gizmodo, March 19, 2019, https://gizmodo.com/its-scary-how-much-personal-data-people-leave-on-used-l-1833383903; Josh Frantz, "Exfiltrating Remaining Private Information from Donated Devices," Rapid7 (blog), March 19, 2019, https://blog.rapid7.com/2019/03/19/buy-one-device-get-data-free-private-information-remains-on-donated-devices/.

59. Frantz, "Exfiltrating Remaining Private Information from Donated Devices."

60. See the 28-minute mark of Sharon Conheady, *"The Future of Social Engineering," presented at DeepSec 2010, Vienna, November 26, 2010,* video, 54:16, https://www.youtube.com/watch?v=bzWRtxA5DCo..

61. Daniel Trottier, "Open Source Intelligence, Social Media and Law Enforcement: Visions, Constraints and Critiques," *European Journal of Cultural Studies* 18, no. 4–5 (August 1, 2015): 533, https://doi.org/10.1177/1367549415577396.

62. Michael Bazzell, *Open Source Intelligence Techniques,* 7th ed., self-published, CreateSpace Independent Publishing Platform, 2019.

63. Ståle Grut, "OSINT Journalism Goes Mainstream," *NiemanLab* (blog), January 3, 2020, https://www.niemanlab.org/2020/01/osint-journalism-goes-mainstream/.

64. Trottier, "Open Source Intelligence, Social Media and Law Enforcement," 533.

65. Mark Andrejevic, *Infoglut: How Too Much Information Is Changing the Way We Think and Know* (New York: Routledge, 2013).

66. Root and Welch, "The Continuing Consumer Study," 7.

67. Long, *No Tech Hacking,* 1.

Chapter 4

1. Christopher Hadnagy, *Social Engineering: The Science of Human Hacking*. 2nd ed. (Indianapolis: Wiley, 2018), 4–6.

2. Edward L. Bernays, *Crystallizing Public Opinion (1923)*, chap. 5.

3. Dual Core's "Trust Me" was the theme song of the Social-Engineer.org podcast from episode 13 on: https://www.social-engineer.org/podcast/episode-013-social-engineering-the-hustle/.

4. Sharon Conheady, *Social Engineering in IT Security: Tools, Tactics, and Techniques* (New York: McGraw-Hill Education, 2014), 120.

5. Christopher Hadnagy, *Social Engineering: The Art of Human Hacking* (Indianapolis: Wiley, 2010).

6. Stefanie Duguay, "Dressing up Tinderella: Interrogating Authenticity Claims on the Mobile Dating App Tinder," *Information, Communication & Society* 20, no. 3 (March 4, 2017): 351–367, https://doi.org/10.1080/1369118X.2016.1168471.

7. Conheady, *Social Engineering in IT Security*, 67.

8. Gavin Watson, Andrew Mason, and Richard Ackroyd, *Social Engineering Penetration Testing: Executing Social Engineering Pen Tests, Assessments and Defense* (Waltham, MA: Elsevier Science, 2014), 265.

9. For example, https://www.youtube.com/watch?v=D7ZZp8XuUTE or https://www.youtube.com/watch?v=I0qpNCg4rMo.

10. Edward L. Bernays, *Propaganda (*New York: H. Liveright, 1928), 77.

11. Thomas H. Bivins, "A Golden Opportunity? Edward Bernays and the Dilemma of Ethics," *American Journalism* 30, no. 4 (January 1, 2013): 511–512, https://doi.org/10.1080/08821127.2013.857981. The emphasis is added, both by us and Bivins.

12. Iris Mostegel, "The Original Influencer," *History Today*, February 6, 2019, https://www.historytoday.com/miscellanies/original-influencer.

13. Susan Henry, "'There Is Nothing in This Profession . . . That a Woman Cannot Do': Doris E. Fleischman and the Beginnings of Public Relations," *American Journalism* 16, no. 2 (April 1, 1999): 93, https://doi.org/10.1080/08821127.1999.10739176.

14. Richard Gunderman, "The Manipulation of the American Mind: Edward Bernays and the Birth of Public Relations," The Conversation, July 9, 2015, http://theconversation.com/the-manipulation-of-the-american-mind-edward-bernays-and-the-birth-of-public-relations-44393.

15. Bernays, *Crystallizing Public Opinion*.

16. Jennifer Jane Marshall, "Clean Cuts: Procter & Gamble's Depression-Era Soap-Carving Contests," *Winterthur Portfolio* 42, no. 1 (2008): 63, https://doi.org/10.1086/528905.

17. Marshall, 63.

18. Marshall, 63.

19. Marshall, 65.

20. Larry Tye, *The Father of Spin: Edward L. Bernays and the Birth of Public Relations* (New York: Henry Holt, 2006), 25.

21. Tye, 24–27.

22. The "Torches of Freedom" campaign is now legendary among historians of public relations, who often claim that it caused women to smoke and smashed a taboo against their smoking in public. However, as Vanessa Murphree argues, the campaign was not as taboo-smashing, nor as successful, as is often claimed. First, she finds that few newspapers reported on the event, and, second, that women had been smoking publicly for years prior to the 1929 event. See Vanessa Murphree, "Edward Bernays's 1929 'Torches of Freedom' March: Myths and Historical Significance," *American Journalism* 32, no. 3 (July 3, 2015): 258–281, https://doi.org/10.1080/08821127.2015.1064681.

23. Murphree, 277.

24. Doris E. Fleischman, "Public Relations and the Consumer," presented at the Fashion Group, New York, October 30, 1935, 4.

25. Scott M. Cutlip, *The Unseen Power: Public Relations, A History* (Hillsdale, NJ: Lawrence Erlbaum, 1994).

26. David Michaels, *The Triumph of Doubt: Dark Money and the Science of Deception* (New York: Oxford University Press, 2020).

27. Michael J. Palenchar and Kathy R. Fitzpatrick, "Secret Persuaders: Ethical and Rhetorical Perspectives on the Use of Public Relations Front Groups," in *Rhetorical and Critical Approaches to Public Relations II*, eds. Robert L. Heath, Elizabeth L. Toth, and Damion Waymer (New York: Routledge, 2009), 272–289.

28. Edward Bernays and Mark Crispin Miller, *Propaganda*, first paperback ed. (Brooklyn: Ig Publishing, 2004), 23.

29. Molly R. Sauter, "Kevin Mitnick, *New York Times*, and the Media's Conception of the Hacker," in *Making Our World: The Hacker and Maker Movements in Context*, eds. Jeremy Hunsinger and Andrew Schrock (New York: Peter Lang, 2019), 21–35.

30. Michael Hess, "Security Tips from a Legendary Hacker," CBS News, December 19, 2011, https://www.cbsnews.com/news/security-tips-from-a-legendary-hacker/;

"Master Hacker Kevin Mitnick Shares His 'Addiction,'" August 21, 2011, NPR, MP3 audio, 7:49, https://www.npr.org/2011/08/21/139677992/master-hacker-kevin-mitnick-shares-his-addiction; Dhanusha Gokulan, "Famous Hacker Proves Your Firm's System Can Be Accessed in an Hour," *Khaleej Times*, April 1, 2019, https://www.khaleejtimes.com/technology/famous-hacker-proves-your-firms-system-can-be-accessed-in-an-hour.

31. Kevin D. Mitnick, "Kevin Mitnick—'I'm The Hacker Who Changed Sides'; First Person," *Financial Times*, May 30, 2009; Jonathan Littman, "Most Wanted in the Mind of Hacker, Kevin Mitnick," *Computerworld*, January 15, 1996, https://www.computerworld.com/article/2531739/most-wanted.html.

32. Kevin D. Mitnick, *The Art of Deception: Controlling the Human Element of Security* (Hoboken: Wiley, 2002).

33. Carla A. Pfeffer, "'I Don't like Passing as a Straight Woman': Queer Negotiations of Identity and Social Group Membership," *American Journal of Sociology* 120, no. 1 (2014): 1–44, https://doi.org/10.1086/677197; Raewyn Connell, "Accountable Conduct: 'Doing Gender' in Transsexual and Political Retrospect," *Gender & Society* 23, no. 1 (2009): 104–111.

34. Connell, "Accountable Conduct," 108.

35. Pfeffer, "'I Don't Like Passing as a Straight Woman,'" 38.

36. Michel Serres, *The Parasite*, trans. Lawrence R. Schehr (Baltimore: Johns Hopkins University Press, 1982).

37. Steven D. Brown, "In Praise of the Parasite: The Dark Organizational Theory of Michel Serres," *Informática Na Educação: Teoria & Prática* 16, no. 1 (2013): 95.

38. Kevin Hetherington and Nick Lee, "Social Order and the Blank Figure," *Environment and Planning D: Society and Space* 18, no. 2 (2000): 170.

39. Sierk Ybema, Tom Keenoy, Cliff Oswick, Armin Beverungen, Nick Ellis, and Ida Sabelis, "Articulating Identities," *Human Relations* 62, no. 3 (March 1, 2009): 306, 315, https://doi.org/10.1177/0018726708101904.

40. Mitnick Security, "The World's Top Security Testing Team," 2019, https://mitnicksecurity.com/.

41. Kevin D. Mitnick and William L. Simon, *Ghost in the Wires: My Adventures as the World's Most Wanted Hacker* (London: Little, Brown, 2012), 5.

42. Mitnick and Simon, 6.

43. Noah McClain, "Caught inside the Black Box: Criminalization, Opaque Technology, and the New York Subway MetroCard," *The Information Society: An International Journal* 35, no. 5 (July 25, 2019): 251–271, https://doi.org/10.1080

/01972243.2019.1644410; Harold Stolper and Jeff Jones, "The Crime of Being Short $2.75: Policing Communities of Color at the Turnstile," New York: Community Service Society, October 2017, https://www.cssny.org/publications/entry/the-crime-of-being-short-2.75.

44. Mitnick and Simon, *Ghost in the Wires*, 6.

45. Raygine DiAquoi, "Symbols in the Strange Fruit Seeds: What 'the Talk' Black Parents Have with Their Sons Tells Us about Racism," *Harvard Educational Review* 87, no. 4 (December 1, 2017): 512–537, https://doi.org/10.17763/1943-5045-87.4.512.

46. Mitnick and Simon, *Ghost in the Wires*, 28.

47. Chris Hadnagy, Robin Dreeke, and Perry Carpenter, "Ep 120—Sizing People Up—LIVE AT DEF CON 27 with Robin Dreeke," August 19, 2019, in Security through Education (podcast), MP3 audio, 01:09:14, https://www.social-engineer.org/podcast/ep-120-sizing-people-up-live-at-def-con-27-with-robin-dreeke/.

48. Mitnick and Simon, *Ghost in the Wires*, 10.

49. Karen Lee Ashcraft, Timothy R. Kuhn, and François Cooren, "Constitutional Amendments: 'Materializing' Organizational Communication," *The Academy of Management Annals* 3, no. 1 (2009): 1–64, https://doi.org/10.1080/19416520903047186.

50. Jennifer L. Gibbs, Dina Nekrassova, Svetlana V. Grushina, and Sally Abdul Wahab, "Reconceptualizing Virtual Teaming from a Constitutive Perspective Review, Redirection, and Research Agenda," *Annals of the International Communication Association* 32, no. 1 (January 2008): 192, https://doi.org/10.1080/23808985.2008.11679078.

51. Mitnick and Simon, *Ghost in the Wires*, 124.

52. Mitnick and Simon, 124–125.

53. Tsutomu Shimomura and John Markoff, *Takedown: The Pursuit and Capture of Kevin Mitnick, America's Most Wanted Computer Outlaw—by the Man Who Did It* (New York: Hyperion, 1996).

54. Joseph D. Won, "Yellowface Minstrelsy: Asian Martial Arts and the American Popular Imaginary," PhD diss. (University of Michigan, ProQuest Dissertations Publishing, 1996), 2, http://search.proquest.com/docview/304250501/?pq-origsite=primo.

55. Won, 3.

56. John Markoff, "Taking a Computer Crime to Heart," *New York Times*, January 28, 1995, https://www.nytimes.com/1995/01/28/business/taking-a-computer-crime-to-heart.html.

57. Kent A. Ono, "'America's' Apple Pie: Baseball, Japan-Bashing, and the Sexual Threat of Economic Miscegenation," in *Out of Bounds: Sports, Media, and the Politics of Identity*, ed. Aaron Baker and Todd Boyd, 81–101 (Bloomington: University of Indiana Press, 1997).

58. Jeff Goodell, "The Samurai and the Cyberthief," *Rolling Stone*, no. 707 (May 4, 1995): 40; This article would be transformed into a book: Jeff Goodell, *The Cyberthief and the Samurai: The True Story of Kevin Mitnick, and the Man Who Hunted Him Down* (New York: Dell, 1996).

59. David Morley and Kevin Robins, "Techno-Orientalism: Japan Panic," chap. 8 in *Spaces of Identity: Global Media, Electronic Landscapes and Cultural Boundaries* (New York: Routledge, 1995); Toshiya Ueno, "Techno-Orientalism and Media-tribalism: On Japanese Animation and Rave Culture," *Third Text* 13, no. 47 (1999): 95–106. https://doi.org/10.1080/09528829908576801.

60. Jonathan Littman, *The Fugitive Game: Online with Kevin Mitnick* (Boston: Little, Brown, 1997), 247.

61. The picture accompanies a story by Steven Levy, "The Case for Hackers," *Newsweek*, February 6, 1995. We have found an archival copy of this photo. While the picture is a bit obnoxious, aping the postmodern aesthetic of 1990s-era *Wired* magazine within the sedate layout of *Newsweek*, Littman's description of the picture is baffling. Littman describes Shimomura as sitting in a "Buddha pose." He might be better described as sitting cross-legged, as he himself describes it in his memoir (Shimomura and Markoff, *Takedown*, 132). Littman refers to him as a "warrior." Shimomura might also reject that framing, since the aesthetic he attributed to himself was as a California rollerblading and skiing bum—indeed, in the photo he's wearing a t-shirt and shorts, and his catchphrase in interviews and in *Takedown* may as well have been "I wanted to get back to skiing" (Goodell, *The Cyberthief and the Samurai*, 316, 321.) And while we've looked at several reprints of the photo Littman is describing, we simply don't see "the sword of a samurai."

62. Littman, *The Fugitive Game*, 257.

63. Littman, 244.

64. "Scenic Drive," Takedown, 1995, http://www.takedown.com/evidence/transcripts/best.html.

65. Mitnick and Simon, *Ghost in the Wires*, 274.

66. Mitnick and Simon, 275.

67. Mitnick and Simon, 276–277.

68. Indeed, many news stories were sure to print Mitnick's acknowledgment of Shimomura when the two met at prearrangement hearing: "Hello, Tsutomu," Mitnick said. "I respect your skills."

69. Jessa Lingel, "Adjusting the Borders: Bisexual Passing and Queer Theory," *Journal of Bisexuality* 9, no. 3–4 (2009): 397, https://doi.org/10.1080/15299971090 3316646.

70. Steven D. Brown, "Michel Serres: Science, Translation and the Logic of the Parasite," *Theory, Culture & Society* 19, no. 3 (2002): 16, https://doi.org/10.1177 /026327602401081503.

71. Karen Lee Ashcraft and Brenda J. Allen, "The Racial Foundation of Organizational Communication," *Communication Theory* 13, no. 1 (2003): 26, https://doi.org /10.1111/j.1468-2885.2003.tb00280.x.

72. Starting around the 43-minute mark of Sharon Conheady, "Social Engineering for Penetration Testers," presented at BruCON Security Conference, Ghent, Belgium, October 3, 2018, video, 1:02:23, https://www.youtube.com/watch?v =oAJ1pNJnJHQ.

73. Sue Curry Jansen, "Semantic Tyranny: How Edward L. Bernays Stole Walter Lippmann's Mojo and Got Away With It and Why It Still Matters," *International Journal of Communication* 7 (April 30, 2013): 1103.

74. In *Public Opinion* (New York: Harcourt, Brace, 1922), Walter Lippmann uses the famous example of the Melting Pot ceremony, put on by Henry Ford, where immigrants would dress in garb stereotypical of their countries of origin, climb into a giant pot, and emerge from the pot "dressed in derby hats, coats, pants, vest, stiff collar and polka-dot tie, . . . and all singing the Star-Spangled Banner."

75. Bernays, *Crystallizing Public Opinion*, chap. 5.

76. Bernays, *Crystallizing Public Opinion*.

77. Bernays.

Chapter 5

1. Kevin D. Mitnick, *The Art of Deception: Controlling the Human Element of Security* (Hoboken: Wiley, 2002).

2. Quoted in Karen Miller Russell and Carl O. Bishop, "Understanding Ivy Lee's Declaration of Principles: U.S. Newspaper and Magazine Coverage of Publicity and Press Agentry, 1865–1904," *Public Relations Review* 35, no. 2 (June 2009): 91, https:// doi.org/10.1016/j.pubrev.2009.01.004.

3. Harry G. Frankfurt, *On Bullshit* (Princeton, NJ: Princeton University Press, 2005): 33.

4. Frankfurt, 55.

5. Frankfurt, 56.

6. E.g., G. A. Cohen, "Deeper into Bullshit," in *The Contours of Agency: Essays on Themes from Harry Frankfurt*, eds. Sarah Buss and Lee Overton (Cambridge, MA: MIT Press, 2002), 321–339; Andreas Stokke and Don Fallis, "Bullshitting, Lying, and Indifference toward Truth," *Ergo, an Open Access Journal of Philosophy* 4, no. 10 (2017), https://doi.org/10.3998/ergo.12405314.0004.010.

7. Stokke and Fallis, "Bullshitting, Lying, and Indifference toward Truth."

8. Stokke and Fallis.

9. Stokke and Fallis.

10. Daniel P. Mears, "The Ubiquity, Functions, and Contexts of Bullshitting," *Journal of Mundane Behavior* 3, no. 2 (June 2002): 245.

11. Chandra Mukerji, "Bullshitting: Road Lore Among Hitchhikers," *Social Problems* 25, no. 3 (February 1, 1978): 242, https://doi.org/10.2307/800062.

12. Mukerji, 246.

13. Mukerji, 246.

14. Mears, "The Ubiquity, Functions, and Contexts of Bullshitting," 242.

15. Frankfurt, *On Bullshit*, 52.

16. Frankfurt, 22.

17. Mukerji, "Bullshitting," 244.

18. Fred Steinbeck, "Verification," *TAP*, November 1983, http://www.textfiles.com/tap/issue.88/tap8804.jpg.

19. Flying Penguin, "Flying Penguin Presents: Bullshitting the Operator," in *The Computer Underground: Computer Hacking, Crashing, Pirating, and Phreaking*, ed. M. Harry (Port Townsend, WA: Loompanics Unlimited, 1985), 87.

20. Flying Penguin, 88.

21. Judas Gerard, "CN/a," *TAP*, October 1982, http://www.textfiles.com/tap/issue.78/tap7803.jpg.

22. Sharp Remob, "Sharp Remob's Guide to Bullshitting the Phone Company Out Of Important Information," archive, Textfiles.com, 1989, http://www.textfiles.com/phreak/soceng.txt.

23. Fred Steinbeck, "Dealing with the Rate & Route Operator," *TAP*, November 1983, http://www.textfiles.com/tap/issue.88/tap8803.jpg.

24. Flying Penguin, "Flying Penguin Presents: Bullshitting the Operator."

25. BIOC Agent 003, "Course in Basic Telecommunications Part II," December 8, 1983, https://textfiles.com/phreak/BIOCAGENT/bioc2.wri.

26. Starting around the 20-minute mark of SN, Chesire Catalyst, and Emmanuel Goldstein, "Hackers on Planet Earth: Social Engineering."

27. The Jammer and Jack the Ripper, "The Official Phreaker's Manual," February 14, 1987. http://textfiles.com/phreak/PHREAKING/manual1.txt.

28. Gerard, "CN/a."

29. Timothy R. Levine, Kim B. Serota, Hillary Shulman, David D. Clare, Hee Sun Park, Allison S. Shaw, Jae Chul Shim, and Jung Hyon Lee, "Sender Demeanor: Individual Differences in Sender Believability Have a Powerful Impact on Deception Detection Judgments," *Human Communication Research* 37, no. 3 (2011): 377–403.

30. Sharp Remob, "Sharp Remob's Guide to Bullshitting the Phone Company Out Of Important Information."

31. The Kid & Co., "How to Get into a C.O." *2600: The Hacker Quarterly* 2, no. 3 (March 1985).

32. David Autovon, "The Automatic Wiretap: Are Your Telephone Conversations Really Private?" *Telephone Electronics Line*, December 1974, 8.

33. Autovon, 8.

34. Autovon, 8. Indeed, in *Exploding the Phone* (New York: Grove, 2014), Phil Lapsley reports that John Draper used verification to spy on one of his (soon-to-be-ex) girlfriends as well as the FBI's field office in San Francisco. In *The Fugitive Game*, Jonathan Littman describes Kevin Mitnick, Lewis De Payne, and Eric Heinz's fascination with SAS, an "automated computerized test system that . . . [can] wiretap anybody's phone or data line." Jonathan Littman, *The Fugitive Game: Online with Kevin Mitnick* (Boston: Little, Brown, 1997), 24.

35. Steinbeck, "Dealing with the Rate & Route Operator."

36. BIOC Agent 003, "Course in Basic Telecommunications Part I," November 15, 1983. http://textfiles.com/phreak/BIOCAGENT/bioc1.wri; BIOC Agent 003, "Course in Basic Telecommunications Part II," December 8, 1983, https://textfiles.com/phreak/BIOCAGENT/bioc2.wri; BIOC Agent 003, "Course in Basic Telecommunications Part IV," April 13, 1984, https://textfiles.com/phreak/BIOCAGENT/bioc2.wri; BIOC Agent 003, "Course in Basic Telecommunications Part II," July 18, 1984, https://textfiles.com/phreak/BIOCAGENT/basicom2.phk; BIOC Agent 003 and Tharrys Ridenow, "Word-Processed Redoing of Bioc Agent 003's Course in Telecommunications," June 18, 1984, http://textfiles.com/phreak/BIOCAGENT/bioc003.001.

37. Mitnick, *The Art of Deception*.

38. Johnny Long, *No Tech Hacking: A Guide to Social Engineering, Dumpster Diving, and Shoulder Surfing*, ed. Scott Pinzon (Rockland, MA: Syngress, 2008), xxi.

39. Christopher Hadnagy, "Social Engineering Code of Ethics," *Security Through Education* (blog), April 4, 2019, https://www.social-engineer.org/framework/general-discussion/social-engineering-code-of-ethics/.

40. For Ivy Lee's stance on "the facts" versus other vehicles for persuasion, such as emotional appeals, see Scott M. Cutlip, *The Unseen Power: Public Relations, A History* (Hillsdale, NJ: Lawrence Erlbaum, 1994), 63. For an exception to our claim that the mass social engineers didn't refer to their work as "bullshit," see John Martin Campbell, *Slinging the Bull in Korea: An Adventure in Psychological Warfare* (Albuquerque: University of New Mexico Press, 2010).

41. Hippolyte Havel, "The Civil War in Colorado," *Mother Earth* 9 (March 1914): 76.

42. Ida M. Tarbell, *The History of the Standard Oil Company* (New York: McClure, Phillips & Co., 1904).

43. Stuart Ewen, *PR! A Social History of Spin* (New York: Basic Books, 1996), 78; Also see Kirk Hallahan, "Ivy Lee and the Rockefellers' Response to the 1913–1914 Colorado Coal Strike," *Journal of Public Relations Research* 14, no. 4 (October 1, 2002): 271, https://doi.org/10.1207/S1532754XJPRR1404_1.

44. Ewen, *PR! A Social History of Spin*.

45. Ray Eldon Hiebert, *Courtier to the Crowd; the Story of Ivy Lee and the Development of Public Relations* (Ames, IA: Iowa State University Press, 1966), 101, http://archive.org/details/courtiertocrowds0000unse.

46. Ewen, *PR! A Social History of Spin*, 78.

47. Hallahan, "Ivy Lee and the Rockefellers' Response to the 1913–1914 Colorado Coal Strike," 302.

48. Ewen, *PR! A Social History of Spin*, 79.

49. Quoted in Frederick Elmore Lumley, *The Propaganda Menace, By Frederick E. Lumley*, (New York: Century, 1933), 77.

50. Quoted in full in Russell and Bishop, "Understanding Ivy Lee's Declaration of Principles," 91.

51. Hallahan, "Ivy Lee and the Rockefellers' Response to the 1913–1914 Colorado Coal Strike."

52. Hallahan, 287.

53. Hallahan, 281.

54. Cutlip, *The Unseen Power*, 120–123.

55. Our argument here echoes that of Cory Wimberly, a scholar of propaganda and public relations. Wimberly argues that what we're calling mass social engineering was less concerned with truth and lies and more concerned with the effects of statements, actions, or events. Statements, actions, or event that effectively shaped public opinion—whether true or not—were valued. Any that did not, were not. See Cory Wimberly, *How Propaganda Became Public Relations*, 162.

56. Hallahan, "Ivy Lee and the Rockefellers' Response to the 1913–1914 Colorado Coal Strike," 269.

57. Cutlip, *The Unseen Power*, 45.

58. Jonathan Auerbach, *Weapons of Democracy: Propaganda, Progressivism, and American Public Opinion* (Baltimore: Johns Hopkins University Press, 2015), 137.

59. Edward L. Bernays, "Torches-031710," SpectorPR, 2010, video, 6:29, https://www.youtube.com/watch?time_continue=299&v=6pyyP2chM8k&feature=emb_logo.

60. Bernays, "Torches-031710."

61. "Easter Sun Finds the Past in Shadow at Modern Parade," *New York Times*, March 1, 1929.

62. Vanessa Murphree, "Edward Bernays's 1929 'Torches of Freedom' March: Myths and Historical Significance," *American Journalism* 32, no. 3 (July 3, 2015): 258–281, https://doi.org/10.1080/08821127.2015.1064681.

63. Susan Henry, "'There Is Nothing in This Profession . . . That a Woman Cannot Do': Doris E. Fleischman and the Beginnings of Public Relations," *American Journalism* 16, no. 2 (April 1, 1999): 85–111, https://doi.org/10.1080/08821127.1999.10739176.; Margot Opdycke Lamme, "Outside the Prickly Nest: Revisiting Doris Fleischman," *American Journalism* 24, no. 3 (July 1, 2007): 85–107, https://doi.org/10.1080/08821127.2007.10678080.

64. Henry, "'There Is Nothing in This Profession . . . That a Woman Cannot Do,'" 100.

65. Frankfurt, *On Bullshit*, 22.

Chapter 6

1. Sharon Conheady, *Social Engineering in IT Security: Tools, Tactics, and Techniques* (New York: McGraw-Hill Education, 2014), 50.

2. Jayson E. Street, "DEFCON 19: Steal Everything, Kill Everyone, Cause Total Financial Ruin! (W Speaker)," Las Vegas, August 2011, https://www.youtube.com/watch?v=JsVtHqICeKE.

3. Rebecca Slayton, "The Paradoxical Authority of the Certified Ethical Hacker," *Limn*, February 14, 2017, https://limn.it/articles/the-paradoxical-authority-of-the-certified-ethical-hacker/; Richard R. Linde, "Operating System Penetration," in *AFIPS '75: Proceedings of the May 19–22, 1975, National Computer Conference and Exposition* (New York: Association for Computing Machinery, 1975), 361–368, https://doi.org/10.1145/1499949.1500018.

4. B. Hebbard, P. Grosso, T. Baldridge, C. Chan, D. Fishman, P. Goshgarian, T. Hilton, J. Hoshen, K. Hoult, G. Huntley, M. Stolarchuk, and L. Warner, "A Penetration Analysis of the Michigan Terminal System," *ACM SIGOPS Operating Systems Review* 14, no. 1 (January 1, 1980): 7–20, https://doi.org/10.1145/850693.850694.

5. John Quann and Peter Belford, "The Hack Attack—Increasing Computer System Awareness of Vulnerability Threats," in *3rd Applying Technology to Systems; Aerospace Computer Security Conference*, (Orlando: American Institute of Aeronautics and Astronautics, 1987), 157, https://doi.org/10.2514/6.1987-3093.

6. Christopher Hadnagy, *Social Engineering: The Science of Human Hacking*, 2nd ed. (Indianapolis: Wiley, 2018), 164.

7. Miriam L. Matteson, Lorien Anderson, and Cynthia Boyden, "'Soft Skills': A Phrase in Search of Meaning," *Portal: Libraries and the Academy* 16, no. 1 (February 18, 2016): 71–88, https://doi.org/10.1353/pla.2016.0009.

8. Arwen Mohun, "Industrial Genders: Home/Factory," in *Gender & Technology: A Reader*, eds. Nina E. Lerman, Ruth Oldenziel, and Arwen P. Mohun (Baltimore: Johns Hopkins University Press, 2003), 153–176; Judy Wajcman, *Feminism Confronts Technology* (University Park, PA: Pennsylvania State University Press, 1991).

9. Tim Jordan and Paul Taylor, "A Sociology of Hackers," *The Sociological Review*, January 25, 2017: 767, https://journals.sagepub.com/doi/pdf/10.1111/1467-954X.00139.

10. Joseph M. Hatfield, "Social Engineering in Cybersecurity: The Evolution of a Concept," *Computers & Security* 73 (March 1, 2018): 104, https://doi.org/10.1016/j.cose.2017.10.008.

11. Ran Almog and Danny Kaplan, "The Nerd and His Discontent: The Seduction Community and the Logic of the Game as a Geeky Solution to the Challenges of Young Masculinity," *Men and Masculinities* 20, no. 1 (April 1, 2017): 30, https://doi.org/10.1177/1097184X15613831.

12. Katie Hafner and John Markoff, *Cyberpunk: Outlaws and Hackers on the Computer Frontier* (New York: Simon & Schuster, 1995); Jonathan Littman, *The Fugitive Game: Online with Kevin Mitnick* (Boston: Little, Brown, 1997).

13. Andrew Stephen King, "Feminism's Flip Side: A Cultural History of the Pickup Artist," *Sexuality & Culture* 22, no. 1 (March 1, 2018): 308, https://doi.org/10.1007/s12119-017-9468-0.

14. Eric C. Hendriks, "Ascetic Hedonism: Self and Sexual Conquest in the Seduction Community," *Cultural Analysis* 11 (2012): 1–14.

15. Hafner and Markoff, *Cyberpunk*, 351–352.

16. Littman, *The Fugitive Game*, 84.

17. Lewis De Payne, "Sensual Access: The High Tech Guide to Seducing Women Using Your Home Computer." Note that this book was self-published some time in the mid-1990s. A copy, mistakenly attributed to Ross Jeffries, is available online at https://www.scribd.com/document/174590270/Ross-Jeffries-Sensual-Access.

18. "How to S.E. Your Sexy Back," Social-Engineer.org, January 8, 2012, https://www.social-engineer.org/podcast/episode-030-how-to-s-e-your-sexy-back/.

19. Jordan Harbinger and Robert Glover, "Dr. Robert Glover | *No More Mr. Nice Guy* (Episode 145)," January 16, 2012, in Art of Charm (podcast), MP3 audio, 1:06:35, https://www.youtube.com/watch?v=EJp5R_6HOA0; Jordan Harbinger and Susan Kuchinskas, "Susan Kuchinskas, *The Chemistry of Connection* (Episode #137)," October 3, 2011, in Art of Charm (podcast), MP3 audio, 48:16, https://www.youtube.com/watch?v=7yNhtQUk9S8; Jordan Harbinger and Marni. "Marni | What Women Think About Confident Men (episode 94)," May 29, 2009, in Art of Charm (podcast), MP3 audio, 45:25, https://www.youtube.com/watch?v=DVXICSQtR0g.

20. Tripp Lanier, "Jordan Harbinger of Pickup Podcast: How to Be Charismatic (episode 094)," April 26, 2010, in The New Man Podcast, MP3 audio, 30:54, http://www.thenewmanpodcast.com/2010/04/jordan-harbinger-charisma/.

21. Chris Hadnagy, Michele Fincher, and Jordan Harbinger, "Ep 094—The Art of Charm Imitates Life," June 12, 2017, in Security through Education (podcast), MP3 audio, 57:10, https://www.social-engineer.org/podcast/ep-094-art-charm-imitates-life/.

22. Derek Rake, "NLP Seduction Patterns, Routines, Phrases & Scripts," Derek Rake (blog), June 29, 2018, https://derekrake.com/blog/nlp-seduction-patterns/. Emphasis in the original. In fact, a common textual habit of the PUAs seems to be excessive use of italic, bold, and capitalized text in their writing.

23. Joseph M. Reagle Jr., *Hacking Life: Systematized Living and Its Discontents* (Cambridge, MA: MIT Press, 2019), chap. 7.

24. Christopher Hadnagy and Paul Ekman, *Unmasking the Social Engineer: The Human Element of Security*, ed. Paul F. Kelly (Indianapolis: Wiley, 2014); Hadnagy, *Social Engineering*, 2018.

25. Christopher Hadnagy, *Social Engineering: The Art of Human Hacking* (Indianapolis: Wiley, 2010).

26. Conheady, *Social Engineering in IT Security*.

27. Susan Headley, "Social Engineering and Psychological Subversion of Trusted Systems," presented at DEF CON, Las Vegas, August 4, 1995.

28. Thomas Ryan and Gabriella Mauch, "Getting in Bed with Robin Sage," Provide Security and BlackHat USA, 2010.

29. Adrianne Jeffries, "Dating Coach Shows How to Get Classified Military Intel Using Social Engineering," The Verge, August 4, 2013, https://www.theverge.com/2013/8/4/4585994/hacking-people-is-easy-a-dating-coach-shows-how-easy-it-is-to-get-classified-intel.

30. Johnny Long, *No Tech Hacking: A Guide to Social Engineering, Dumpster Diving, and Shoulder Surfing*, ed. Scott Pinzon (Rockland, MA: Syngress, 2008), 16.

31. Mike Winnet and Jenny Radcliffe, "Lies, Cons & People Hacking with Jenny Radcliffe (episode #5)," July 12, 2019, in *Not Another D*ckhead with a Podcast*, video, 50:45, https://www.youtube.com/watch?v=qbKrK753wn0.

32. For a discussion of how pickup artists are advised against monogamy, see Hendriks, "Ascetic Hedonism," 9–11.

33. "Penetration Tester Job Description," JobHero, accessed August 16, 2020, https://www.jobhero.com/job-description/examples/information-technology/penetration-tester.

34. Occupational Outlook Handbook, "Information Security Analysts," Bureau of Labor Statistics, April 9, 2021, https://www.bls.gov/ooh/computer-and-information-technology/information-security-analysts.htm.

35. "2-Day Social Engineering Training Course Outline" Mitnick Security Consulting, 2005; "Program Details » MS in Cybersecurity Program Overview," The University of Arizona, accessed March 13, 2020, https://cybersecurity.arizona.edu/program/; Social-Engineer—Professional Social Engineering Training and Services, "2-Day Social Engineering Bootcamp," accessed March 13, 2020.

36. Occupational Outlook Handbook, "Information Security Analysts."

37. Molly R. Sauter, "Kevin Mitnick, the *New York Times*, and the Media's Conception of the Hacker," in *Making Our World: The Hacker and Maker Movements in Context*, eds. Jeremy Hunsinger and Andrew Schrock (New York: Peter Lang, 2019), 28.

38. Henry M. Kluepfel, "Foiling the Wiley Hacker: More than Analysis and Containment," in *Proceedings. International Carnahan Conference on Security Technology*, 16, 1989, https://doi.org/10.1109/CCST.1989.751947.

39. "The Electronic Delinquents," ABC, April 22, 1982.

40. "Testimony of Susan Headley, Tujunga, Calif.," in *Computer Security in the Federal Government and the Private Sector: Hearings before the Subcommittee on Oversight of Government Management of the Committee on Governmental Affairs, United States Senate, Ninety-Eighth Congress, First Session, October 25 and 26, 1983*, 22–29, Washington, DC: Government Printing Office, 1983; Hafner and Markoff, *Cyberpunk*.

41. Mary Thornton, "Even Checking Lip Prints Wouldn't Offer Absolute Protection," *Washington Post*, May 22, 1984; Megan Rosenfeld, "At Surveillance Expo, Sneak Peeks at the Sweet Spy and Buy," *Washington Post*, December 18, 1989, https://www.washingtonpost.com/archive/business/1989/12/18/at-surveillance-expo-sneak-peeks-at-the-sweet-spy-and-buy/4b130402-43b1-4942-ada7-df34de8e51de/.

42. "Kevin Mitnick: Cyber Thief; Computer Hacker Is Caught and Sent to Prison for Invading Company Computers in What He Says Was Just a Challenge to Him," CBS, January 23, 2000; Kevin Mitnick, "Kevin Mitnick's Testimony to Senate Homeland Security and Government Affairs Committee," *Cyber Attack: Is the Government Safe?* Governmental Affairs Committee (March 2, 2000), https://www.hsgac.senate.gov/imo/media/doc/mitnick.pdf.

43. Stephen Lynch, "Kevin Mitnick Reboots His Life // For the World's Most Notorious Computer Hacker, the Hardest Thing to Crack Is His Own Reputation," *Orange County Register*, November 8, 2001; Brian Fonesca, "Kevin Mitnick, Hacker Extraordinaire, Speaks out on Security in Today's Internet Age," *InfoWorld*, December 1, 2000.

44. "2-Day Social Engineering Training Course Outline," Mitnick Security Consulting, 2005.

45. "The SEORG Book List," Security Through Education, August 14, 2019, https://www.social-engineer.org/resources/seorg-book-list/.

46. Edward Bernays and Mark Crispin Miller, *Propaganda*, first paperback ed. (Brooklyn: Ig Publishing, 2004); Edward L. Bernays and Howard Walden Cutler, *The Engineering of Consent* (Norman: University of Oklahoma Press, 1955); Robert B. Cialdini, *Influence: The Psychology of Persuasion* (New York: Quill, 1985); Dale Carnegie, *How to Win Friends & Influence People*, 80th anniversary ed. (New York: Simon & Schuster, 2017).

47. Joe Navarro and Marvin Karlins, *What Every BODY Is Saying: An Ex-FBI Agent's Guide to Speed-Reading People* (New York: William Morrow, 2008); Paul Ekman, *Emotions Revealed: Understanding Faces and Feelings* (London: Orion, 2004); Paul Ekman and Wallace V. Friesen, *Unmasking the Face: A Guide to Recognizing Emotions from Facial Expressions* (Los Angeles: Malor, 2003); Ekman and Friesen.

48. Robin K. Dreeke, *It's Not All about "Me": The Top Ten Techniques for Building Quick Rapport with Anyone*, self-published (People Formula, 2011); Daniel Kahneman, *Thinking, Fast and Slow* (New York: Farrar, Straus & Giroux, 2011); Ellen J. Langer, *Mindfulness*, 25th anniversary ed. (Boston: Da Capo Lifelong Books, 2014); Ellen J. Langer, *The Power of Mindful Learning*, reprint ed. (Boston: Da Capo Lifelong Books, 2016).

49. Long, *No Tech Hacking*; Kevin D. Mitnick, *The Art of Deception: Controlling the Human Element of Security* (Hoboken: Wiley, 2002); Kevin D. Mitnick and William L. Simon, *Ghost in the Wires: My Adventures as the World's Most Wanted Hacker* (London: Little, Brown, 2012); Hadnagy, *Social Engineering*, 2010; Hadnagy, *Social Engineering*, 2018.

50. Conheady, *Social Engineering in IT Security*.

51. Conheady, 163.

52. Street, "DEFCON 19"; Headley, "Social Engineering and Psychological Subversion of Trusted Systems."

53. See in particular the 23-minute mark of Street, "DEFCON 19."

54. Slayton, "The Paradoxical Authority of the Certified Ethical Hacker."

55. Richard Butsch, *The Citizen Audience: Crowds, Publics, and Individuals* (London: Routledge, 2007), 11; Paul Starr, *The Creation of the Media: Political Origins of Modern Communication* (New York: Basic Books, 2005).

56. James W. Carey, *Communication as Culture: Essays on Media and Society* (New York: Routledge, 1989), 6–8.

57. Butsch, *The Citizen Audience: Crowds, Publics, and Individuals*, 81.

58. Ray Stannard Baker, "Railroads on Trial," *McClure's Magazine*, March 1906, 537.

59. Quoted in Jonathan Auerbach, *Weapons of Democracy: Propaganda, Progressivism, and American Public Opinion* (Baltimore: Johns Hopkins University Press, 2015), 70.

60. Auerbach, 13.

61. See for example Leo Jay Margolin, *Paper Bullets: A Brief Story of Psychological Warfare in World War II* (New York: Froben Press, 1946).

62. Edward L. Bernays, "The Marketing of National Policies: A Study of War Propaganda," *Journal of Marketing* 6, no. 3 (1942): 239, https://doi.org/10.1177/002224294200600303.

63. Edward L. Bernays, "The Engineering of Consent," *The ANNALS of the American Academy of Political and Social Science* 250, no. 1 (March 1947), 119–120, https://doi.org/10.1177/000271624725000116.

64. Ghislain Thibault, "Needles and Bullets: Media Theory, Medicine, and Propaganda, 1910–1940," in *Endemic: Essays in Contagion Theory*, eds. Kari Nixon and Lorenzo Servitje (London: Palgrave Macmillan, 2016), 67–92, https://doi.org/10.1057/978-1-137-52141-5_4. We should note that there has been a decades-long debate among professional scholars of communication about the degree to which early communication theorists subscribed to a "magic bullet" or "hypodermic needle" theory that posited the ability to achieve direct effects on audiences. The debate seems to have been resolved in favor of the argument that no formal theory identified as "magic bullet" or "hypodermic needle" ever existed. While there may not have been a formal theory, there was certainly significant use of penetration metaphors to talk about the potential for mass persuasion to achieve direct effects on audiences. For further discussion of this debate, see: Jeffrey L. Bineham, "A Historical Account of the Hypodermic Model in Mass Communication," *Communication Monographs* 55, no. 3 (September 1988): 230–246, https://doi.org/10.1080/03637758809376169; J. Michael Sproule, "Progressive Propaganda Critics and the Magic Bullet Myth," *Critical Studies in Mass Communication* 6, no. 3 (September 1989): 225–246, https://doi.org/10.1080/15295038909366750; Butsch, *The Citizen Audience*, 16, 118.

65. Christopher Simpson, *Science of Coercion: Communication Research and Psychological Warfare, 1945–1960* (New York: Oxford University Press, 1994), 62; Also see Butsch, *The Citizen Audience*, 123–124.

66. For an early example, see the discussion of news clips about AT&T in Scott M. Cutlip, *The Unseen Power: Public Relations, A History* (Hillsdale, NJ: Lawrence Erlbaum, 1994), 18–19.

67. Larry Tye, *The Father of Spin: Edward L. Bernays and the Birth of Public Relations* (New York: Henry Holt, 2006), 40.

68. Tye, 40.

69. Edward L. Bernays, *Biography of an Idea: Memoirs of Public Relations Counsel* (New York: Simon & Schuster), 1965, http://archive.org/details/biographyofideam00bern; Margot Opdycke Lamme, "Outside the Prickly Nest: Revisiting Doris Fleischman," *American Journalism* 24, no. 3 (July 1, 2007): 85–107, https://doi.org/10.1080/08821127.2007.10678080.

70. Kirk Hallahan, "Ivy Lee and the Rockefellers' Response to the 1913–1914 Colorado Coal Strike," *Journal of Public Relations Research* 14, no. 4 (October 1, 2002): 271, https://doi.org/10.1207/S1532754XJPRR1404_1.

71. Simpson, *Science of Coercion*, 29; For an example of the transition from simple clip counting to random sampling, see Nathan Maccoby, Freddie O. Sabghir, and Bryant Cushing, "A Method for the Analysis of the News Coverage of Industry,"

Public Opinion Quarterly 14, no. 4 (January 1, 1950): 753–758, https://doi.org/10.1086/266253.

72. Robert O. Carlson, "The Use of Public Relations Research by Large Corporations," *Public Opinion Quarterly* 21, no. 3 (1957): 347, https://doi.org/10.1086/266726.

73. Bernays, "The Marketing of National Policies," 240.

74. Ivy Ledbetter Lee, *Human Nature and the Railroads,* (E. S. Nash, 1915), 22, http://archive.org/details/humannatureandr01leegoog.

75. Auerbach, *Weapons of Democracy*, 13.

76. Auerbach, 168.

77. Auerbach, 168; Also see Bruce Lannes Smith, "Propaganda Analysis and the Science of Democracy," *Public Opinion Quarterly* 5, no. 2 (June 1941): 250–259, https://doi.org/10.1086/265490.

78. Edward L. Bernays, *The Engineering of Consent* (Philadelphia 1947), 114–115.

79. Lee, *Human Nature and the Railroads*, 23.

80. Bernays, "The Engineering of Consent," 113; Edward L. Bernays, "Human Engineering and Social Adjustment," *ETC: A Review of General Semantics* 74, no. 3–4 (July 2017): 351.

81. Simpson, *Science of Coercion*; Mark Solovey, "Science and the State During the Cold War: Blurred Boundaries and a Contested Legacy," *Social Studies of Science* 31, no. 2 (2001): 165–170, https://doi.org/10.1177/030631270103100200; Mark Solovey, "Project Camelot and the 1960s Epistemological Revolution: Rethinking the Politics-Patronage-Social Science Nexus," *Social Studies of Science* 31, no. 2 (2001): 171–206, https://doi.org/10.1177/030631270103100203.

82. Armand Mattelart, *Networking the World, 1794–2000*, English language ed., trans. Liz Carey-Libbrecht and James A. Cohen (Minneapolis: University of Minnesota Press, 2000): 55–56.

83. Hadnagy, *Social Engineering*, 2018, 2.

84. Conheady, *Social Engineering in IT Security*, 1.

85. Mitnick, *The Art of Deception*.

86. Hadnagy, *Social Engineering*, 2010.

87. Hadnagy.

88. Social-Engineer, "The Human Hacking Conference 2021: Plan to Be Amazed!" Security Through Education (blog), June 22, 2020, https://www.social-engineer.org/social-engineering/the-human-hacking-conference-2021-plan-to-be-amazed/.

89. Auerbach, *Weapons of Democracy*, 3.

90. Slayton, "The Paradoxical Authority of the Certified Ethical Hacker."

Chapter 7

1. Christopher Wylie, *Mindf*ck: Cambridge Analytica and the Plot to Break America* (New York: Random House, 2019), 16.

2. Edward L. Bernays, "Human Engineering and Social Adjustment," *ETC: A Review of General Semantics* 74, no. 3–4 (July 2017): 346–351.

3. Patrick B. O'Sullivan and Caleb T. Carr, "Masspersonal Communication: A Model Bridging the Mass-Interpersonal Divide," *New Media & Society* 20, no. 3 (March 1, 2018): 1164, https://doi.org/10.1177/1461444816686104.

4. This is not to suggest that there aren't other communicative practices that are also manipulative, unethical, and even potentially damaging. Again, our goal here is to explicate the emergent form of such communication: masspersonal social engineering.

5. Mél Hogan, "The Archive as Dumpster," *Pivot: A Journal of Interdisciplinary Studies & Thought* 4, no. 1 (2015), https://doi.org/10.25071/2369-7326.39565.

6. Raphael Satter, Jeff Donn, and Chad Day, "Inside Story: How Russians Hacked the Democrats' Emails," AP News, November 4, 2017, https://apnews.com/dea73efc01594839957c3c9a6c962b8a.

7. United States Department of Justice, "United States vs. Internet Research Agency, et al.," February 16, 2018, 16, https://www.justice.gov/file/1035477/download.

8. "United States vs. Internet Research Agency et al.," 13.

9. "United States vs. Internet Research Agency et al.," 5.

10. Nanna Bonde Thylstrup, "Data Out of Place: Toxic Traces and the Politics of Recycling," *Big Data & Society* 6, no. 2 (July 1, 2019): 4, https://doi.org/10.1177/2053951719875479.

11. For a discussion of the digital dumpster, see Hogan, "The Archive as Dumpster."

12. Matthew Rosenberg, Nicholas Confessore, and Carole Cadwalladr, "How Trump Consultants Exploited the Facebook Data of Millions," *New York Times*, March 17, 2018, https://www.nytimes.com/2018/03/17/us/politics/cambridge-analytica-trump-campaign.html.

13. Carole Cadwalladr and Emma Graham-Harrison, "Revealed: 50 Million Facebook Profiles Harvested for Cambridge Analytica in Major Data Breach," *Guardian*, March 17, 2018, https://www.theguardian.com/news/2018/mar/17/cambridge

-analytica-facebook-influence-us-election; Rosenberg, Confessore, and Cadwalladr, "How Trump Consultants Exploited the Facebook Data of Millions."

14. Rosenberg, Confessore, and Cadwalladr, "How Trump Consultants Exploited the Facebook Data of Millions."

15. Rosenberg, Confessore, and Cadwalladr.

16. CBS This Morning, "Former Cambridge Analytica Employee Says 'There Were More' Apps That Collected User Data," CBS, June 27, 2018, https://www.cbsnews.com/news/cambridge-analytica-brittany-kaiser-speaks-out-data-collection-facebook/; Brittany Kaiser, *Targeted: The Cambridge Analytica Whistleblower's Inside Story of How Big Data, Trump, and Facebook Broke Democracy and How It Can Happen Again* (New York: HarperCollins, 2019), 78.

17. Wylie, *Mindf*ck*, 113.

18. Kaiser, *Targeted*, 77–78.

19. Kaiser, 98.

20. Thylstrup, "Data Out of Place," 4. Thylstrup is making a good point, but it is important to remember that mining, like many extractive industries, also produces a lot of harmful waste.

21. Indeed, this is a point made by Fiona Hill, "Opinion | The Biggest Risk to This Election Is Not Russia. It's Us," *New York Times*, October 7, 2020, https://www.nytimes.com/2020/10/07/opinion/trump-russia-election-interference.html.

22. "Getting 2 Equal: United Not Divided," State of Black America. New York: National Urban League, 2019, http://soba.iamempowered.com/2019-report, 10.

23. Philip N. Howard, Bharath Ganesh, Dimitra Liotsiou, John Kelly, and Camille François, "The IRA, Social Media and Political Polarization in the United States, 2012–2018" (Oxford: Computational Propaganda Research Project, 2018), 7.

24. Robert W. Gehl and Maria Bakardjieva, "Socialbots and Their Friends," in *Socialbots and Their Friends: Digital Media and the Automation of Sociality*, eds. Robert W. Gehl and Maria Bakardjieva, (London: Routledge, 2016), 1–16.

25. David M. Beskow and Kathleen M. Carley, "Social Cybersecurity: An Emerging National Security Requirement," *Military Review: The Professional Journal of the U.S. Army* (April 2019): 124.

26. Robert Mueller, "Report on the Investigation into Russian Interference in the 2016 Presidential Election," Washington, DC: Department of Justice, March 2019, https://cdn.cnn.com/cnn/2019/images/04/18/mueller-report-searchable.pdf, 28.

27. Beskow and Carley, "Social Cybersecurity: An Emerging National Security Requirement," 125.

28. Damian J. Ruck, Natalie M. Rice, Joshua Borycz, and R. Alexander Bentley, "Internet Research Agency Twitter Activity Predicted 2016 U.S. Election Polls," *First Monday* 24, no. 7 (June 30, 2019), https://doi.org/10.5210/fm.v24i7.10107.

29. Renee DiResta, Kris Shaffer, Becky Ruppel, David Sullivan, Robert Matney, Ryan Fox, Jonathan Albright, and Ben Johnson, "The Tactics & Tropes of the Internet Research Agency," New Knowledge [now Yonder], 2018, 13.

30. DiResta et al., 20.

31. DiResta et al., 44; Also see Michael Bossetta, "The Weaponization of Social Media: Spear Phishing and Cyberattacks on Democracy," *Journal of International Affairs* 70, no. 1.5 (2018): 101.

32. Carole Cadwalladr and Mark Townsend, "Revealed: The Ties That Bound Vote Leave's Data Firm to Controversial Cambridge Analytica," *Guardian*, March 24, 2018, https://www.theguardian.com/uk-news/2018/mar/24/aggregateiq-data-firm-link-raises-leave-group-questions; Harry Davies, "Ted Cruz Campaign Using Firm That Harvested Data on Millions of Unwitting Facebook Users," *Guardian*, December 11, 2015, https://www.theguardian.com/us-news/2015/dec/11/senator-ted-cruz-president-campaign-facebook-user-data; Carole Cadwalladr, "'I Made Steve Bannon's Psychological Warfare Tool': Meet the Data War Whistleblower," *Guardian*, March 18, 2018, https://www.theguardian.com/news/2018/mar/17/data-war-whistleblower-christopher-wylie-faceook-nix-bannon-trump.

33. CBS This Morning, "Former Cambridge Analytica Employee Says 'There Were More' Apps That Collected User Data."

34. "Cambridge Analytica Uncovered: Secret Filming Reveals Election Tricks," *Channel 4 News*, 2018, video, 19:12, https://www.youtube.com/watch?v=mpbeOCKZFfQ.

35. "Cambridge Analytica Uncovered."

36. Alexander Nix, "From Mad Men to Math Men," presented at OMR Festival 2017, Hamburg, March 2017, video, 30:17, https://www.youtube.com/watch?v=6bG5ps5KdDo.

37. Cadwalladr, "'I Made Steve Bannon's Psychological Warfare Tool'"; Also see Ellen Barry, "Long Before Cambridge Analytica, a Belief in the 'Power of the Subliminal' (Published 2018)," *New York Times*, April 20, 2018, https://www.nytimes.com/2018/04/20/world/europe/oakes-scl-cambridge-analytica-trump.html.

38. "Cambridge Analytica Uncovered." For more analysis, see Emma Briant's forthcoming book *Propaganda Machine: Inside Cambridge Analytica and the Digital Influence Industry* (New York: Bloomsbury, 2021).

39. Darren L. Linvill, Brandon C. Boatwright, Will J. Grant, and Patrick L. Warren, "'The Russians Are Hacking My Brain!' Investigating Russia's Internet Research Agency Twitter Tactics During the 2016 United States Presidential Campaign," *Computers in Human Behavior* 99 (October 1, 2019): 292–300, https://doi.org/10.1016/j.chb.2019.05.027.

40. DiResta et al., "The Tactics & Tropes of the Internet Research Agency," 55.

41. Philip N. Howard, Bharath Ganesh, Dimitra Liotsiou, John Kelly, and Camille François, "The IRA, Social Media and Political Polarization in the United States, 2012–2018 (Appendices)" (Oxford: Computational Propaganda Research Project, 2018), 7–8.

42. Harry G. Frankfurt, *On Bullshit* (Princeton, NJ: Princeton University Press, 2005), 22.

43. Andrew Dawson and Martin Innes, "How Russia's Internet Research Agency Built Its Disinformation Campaign," *The Political Quarterly* 90, no. 2 (2019): 247, https://doi.org/10.1111/1467-923X.12690.

44. DiResta et al., "The Tactics & Tropes of the Internet Research Agency."

45. DiResta et al., 45.

46. DiResta et al., 13.

47. Dawson and Innes, "How Russia's Internet Research Agency Built Its Disinformation Campaign," 250.

48. Dmitry Volchek and Daisy Sindelar, "One Professional Russian Troll Tells All," Radio Free Europe/Radio Liberty, March 25, 2015, https://www.rferl.org/a/how-to-guide-russian-trolling-trolls/26919999.html.

49. Wylie, *Mindf*ck*, 121.

50. See the 18-minute mark of Molly Schweickert, "Cambridge Analytica explains how the Trump campaign worked," *D3con: The Future of Digital Advertising, 2017*, video, 40:02, https://www.youtube.com/watch?v=bB2BJjMNXpA.

51. Glenn Kessler, "Analysis | Foundation Faceoff: The Trump Foundation vs. the Clinton Foundation," *Washington Post*, June 27, 2018, https://www.washingtonpost.com/news/fact-checker/wp/2018/06/27/foundation-face-off-the-trump-foundation-versus-the-clinton-foundation/; "Clinton Foundation Archives," *FactCheck.org*, 2020, https://www.factcheck.org/issue/clinton-foundation/.

52. "Cambridge Analytica Uncovered."

53. Karim Amer and Jehane Noujaim, dirs., *The Great Hack,* documentary, The Othrs, 2019, https://www.netflix.com/title/80117542.

54. Wylie, *Mindf*ck*, 178.

55. Wylie, 47.

56. Wylie, 121–122.

57. Wylie, 132 and 66.

58. Wylie, 178.

59. Wylie, 71.

60. Wylie, 85.

61. Peter Pomerantsev, *Nothing Is True and Everything Is Possible: The Surreal Heart of the New Russia* (New York: PublicAffairs, 2014); Christopher Paul and Miriam Matthews, "The Russian 'Firehose of Falsehood' Propaganda Model: Why It Might Work and Options to Counter It" (Santa Monica, CA: RAND Corporation, 2016), https://doi.org/10.7249/PE198; Timothy Snyder, *The Road to Unfreedom: Russia, Europe, America*, reprint ed. (New York: Tim Duggan, 2019).

62. Wylie, 110.

63. "Cambridge Analytica Uncovered."

64. "Cambridge Analytica Uncovered."

65. Nix, "From Mad Men to Math Men."

66. Wylie, *Mindf*ck*, 63 and 93.

67. Dawson and Innes, "How Russia's Internet Research Agency Built Its Disinformation Campaign," 246.

68. "United States vs. Internet Research Agency et al.," 27–30.

69. "Russian Active Measures Campaigns and Interference in the 2016 U.S. Election. Volume 2: Russia's Use of Social Media with Additional Views" (Washington, DC: US Senate Committee on Intelligence, 2019), 39, https://www.intelligence.senate.gov/sites/default/files/documents/Report_Volume2.pdf.

70. Nina Jankowicz, "How an Anti-Trump Flash Mob Found Itself in the Middle of Russian Meddling," POLITICO, July 5, 2020, https://www.politico.com/news/magazine/2020/07/05/how-an-anti-trump-flash-mob-found-itself-in-the-middle-of-russian-meddling-348729.

71. "Russian Active Measures Campaigns and Interference in the 2016 U.S. Election."

72. "Assessing Russian Activities and Intentions in Recent US Elections" (Washington, DC: Office of the Director of National Intelligence, 2017), 5, https://www.dni.gov/files/documents/ICA_2017_01.pdf.

73. Nix, "From Mad Men to Math Men."

74. "Cambridge Analytica Uncovered."

75. Nix, "From Mad Men to Math Men."

76. Kaiser, *Targeted*, 93.

77. Kaiser, 47.

78. Kaiser, 12–13.

79. Wylie, *Mindf*ck*, 55.

80. Schweickert, "Cambridge Analytica explains how the Trump campaign worked"; Nix, "From Mad Men to Math Men."

81. "US Report Finds No Direct Foreign Interference in 2018 Vote," *AP News*, December 21, 2018, https://apnews.com/article/cd2618aaeb6040c5b57fb301361c76fd

82. Victoria Clark, Mikhaila Fogel, Susan Hennessey, Quinta Jurecic, Matthew Kahn, and Benjamin Wittes, "Russian Electoral Interference: 2018 Midterms Edition," *Lawfare* (blog), October 19, 2018, https://www.lawfareblog.com/russian-electoral-interference-2018-midterms-edition; Josh Gerstein, "U.S. Brings First Charge for Meddling in 2018 Midterm Elections," *POLITICO*, October 19, 2018, https://politi.co/2Ajbubq; Adam Goldman, "Justice Dept. Accuses Russians of Interfering in Midterm Elections," *New York Times*, October 19, 2018, https://www.nytimes.com/2018/10/19/us/politics/russia-interference-midterm-elections.html; "Joint Statement from the ODNI, DOJ, FBI and DHS: Combating Foreign Influence in U.S. Elections," Washington, DC: Director of National Intelligence, October 19, 2018, https://www.dni.gov/index.php/newsroom/press-releases/item/1915-joint-statement-from-the-odni-doj-fbi-and-dhs-combating-foreign-influence-in-u-s-elections; "Russian National Charged with Interfering in U.S. Political System," US Department of Justice, October 19, 2018, https://www.justice.gov/opa/pr/russian-national-charged-interfering-us-political-system.

83. Joshua Geltzer, "Don't Be Fooled: There *Was* Election Interference in 2018," *Just Security* (blog), November 7, 2018, https://www.justsecurity.org/61372/dont-fooled-was-election-interference-2018/; Ellen Nakashima, "U.S. Cyber Command Operation Disrupted Internet Access of Russian Troll Factory on Day of 2018 Midterms," *Washington Post*, February 27, 2019, https://www.washingtonpost.com/world/national-security/us-cyber-command-operation-disrupted-internet-access-of-russian-troll-factory-on-day-of-2018-midterms/2019/02/26/1827fc9e-36d6-11e9-af5b-b51b7ff322e9_story.html.

84. "Treasury Sanctions Russia-Linked Election Interference Actors," US Department of the Treasury, September 10, 2020, https://home.treasury.gov/news/press-releases/sm1118. This announcement also included sanctions on three more alleged members of the IRA.

85. Michael R. Pompeo, "Sanctioning Russia-Linked Disinformation Network for Its Involvement in Attempts to Influence U.S. Election," *US Department of State* (blog), January 11, 2021, https://2017-2021.state.gov/sanctioning-russia-linked-disinformation-network-for-its-involvement-in-attempts-to-influence-u-s-election/index.html; "Treasury Takes Further Action Against Russian-Linked Actors," US Department of the Treasury, January 11, 2021, https://home.treasury.gov/news/press-releases/sm1232.

86. "Statement by NCSC Director William Evanina: Election Threat Update for the American Public," Office of the Director of National Intelligence, August 7, 2020, https://www.dni.gov/index.php/newsroom/press-releases/item/2139-statement-by-ncsc-director-william-evanina-election-threat-update-for-the-american-public.

87. Thomas Rid, "Insisting That the Hunter Biden Laptop Is Fake Is a Trap. So Is Insisting That It's Real," *Washington Post*, October 24, 2020, https://www.washingtonpost.com/outlook/2020/10/24/hunter-biden-laptop-disinformation/.

88. Ellen Nakashima and David L. Stern, "U.S. Sanctions Ukrainians Involved in Russia-Linked Campaign Promoted by Giuliani to Smear Biden," *Washington Post*, January 11, 2021, https://www.washingtonpost.com/national-security/ukranians-sanctions-giuliani-election-interference/2021/01/11/0c447aea-5436-11eb-a08b-f1381ef3d207_story.html.

89. "Treasury Sanctions Russia-Linked Election Interference Actors."

90. Alan Cullison, Rebecca Ballhaus, and Dustin Volz, "Trump Repeatedly Pressed Ukraine President to Investigate Biden's Son," *Wall Street Journal*, September 21, 2019, https://www.wsj.com/articles/trump-defends-conversation-with-ukraine-leader-11568993176.

91. Nakashima and Stern, "U.S. Sanctions Ukrainians Involved in Russia-Linked Campaign Promoted by Giuliani to Smear Biden."

92. Shannon Bond, "Facebook And Twitter Limit Sharing 'New York Post' Story About Joe Biden," *NPR.org*, October 14, 2020, https://www.npr.org/2020/10/14/923766097/facebook-and-twitter-limit-sharing-new-york-post-story-about-joe-biden.

93. "Statement by NCSC Director William Evanina."

94. "Alert Number ME-000138-TT: Indicators of Compromise Pertaining to Iranian Interference in the 2020 US Presidential Election," FBI Flash, October 29, 2020, https://www.ic3.gov/Media/News/2020/201030.pdf.

95. Steven Melendez, "Florida Democrats Get Threatening Emails Demanding That They Vote for Trump," *Fast Company*, October 20, 2020, https://www.fastcompany.com/90566468/florida-democrats-get-threatening-emails-demanding-that-they-vote-for-trump.

96. Cybersecurity & Infrastructure Security Agency, "Alert (AA20–304A): Iranian Advanced Persistent Threat Actor Identified Obtaining Voter Registration Data," October 30, 2020, https://us-cert.cisa.gov/ncas/alerts/aa20-304a

97. Ellen Nakashima, Amy Gardner, Isaac Stanley-Becker, and Craig Timberg, "U.S. Government Concludes Iran Was behind Threatening Emails Sent to Democrats," *Washington Post*, October 22, 2020, https://www.washingtonpost.com/technology/2020/10/20/proud-boys-emails-florida/.

98. Ellen Nakashima, Amy Gardner, and Aaron C. Davis, "FBI Links Iran to Online Hit List Targeting Top Officials Who've Refuted Trump's Election Fraud Claims," *Washington Post*, December 22, 2020, https://www.washingtonpost.com/national-security/iran-election-fraud-violence/2020/12/22/4a28e9ba-44a8-11eb-a277-49a6d1f9dff1_story.html.

99. "Alert Number ME-000138-TT."

100. Nakashima, Gardner, and Davis, "FBI Links Iran to Online Hit List Targeting Top Officials Who've Refuted Trump's Election Fraud Claims."

101. Nakashima et al., "U.S. Government Concludes Iran Was behind Threatening Emails Sent to Democrats."

102. "Alert Number ME-000138-TT"; Cybersecurity & Infrastructure Security Agency, "Alert (AA20–304A)."

103. Stanford Internet Observatory, *Reply-Guys Go Hunting: An Investigation into a U.S. Astroturfing Operation on Facebook, Twitter, and Instagram* (Palo Alto, CA: Stanford University, 2020), https://cyber.fsi.stanford.edu/io/news/oct-2020-fb-rally-forge.

104. Isaac Stanley-Becker, "Facebook Bans Marketing Firm Running 'Troll Farm' for Pro-Trump Youth Group," *Washington Post*, October 8, 2020, https://www.washingtonpost.com/technology/2020/10/08/facebook-bans-media-consultancy-running-troll-farm-pro-trump-youth-group/.

105. Stanford Internet Observatory, *Reply-Guys Go Hunting*, 6.

106. Stanford Internet Observatory, 4.

107. Scott Shane and Alan Blinder, "Secret Experiment in Alabama Senate Race Imitated Russian Tactics," *New York Times*, December 20, 2018, https://www.nytimes.com/2018/12/19/us/alabama-senate-roy-jones-russia.html; Casey Newton, "Democrats Ran Influence Campaigns in at Least Three States During the Midterm Elections," *The Verge*, January 8, 2019, https://www.theverge.com/2019/1/8/18173027/democrats-misinformation-reid-hoffman-alabama-new-knowledge-influence-campaigns; Craig Timberg, Tony Romm, Aaron C. Davis, and Elizabeth Dwoskin, "Secret Campaign to Use Russian-Inspired Tactics in 2017 Ala. Election Stirs Anxiety

for Democrats," *Washington Post*, January 6, 2019, https://www.washingtonpost.com/business/technology/secret-campaign-to-use-russian-inspired-tactics-in-2017-alabama-election-stirs-anxiety-for-democrats/2019/01/06/58803f26-0400-11e9-8186-4ec26a485713_story.html; Scott Shane and Alan Blinder, "Democrats Faked Online Push to Outlaw Alcohol in Alabama Race," *New York Times*, January 7, 2019, https://www.nytimes.com/2019/01/07/us/politics/alabama-senate-facebook-roy-moore.html.

108. Shane and Blinder, "Secret Experiment in Alabama Senate Race Imitated Russian Tactics."

109. Shane and Blinder, "Democrats Faked Online Push to Outlaw Alcohol in Alabama Race."

110. Timberg et al., "Secret Campaign to Use Russian-Inspired Tactics in 2017 Ala. Election Stirs Anxiety for Democrats."

111. Shane and Blinder, "Democrats Faked Online Push to Outlaw Alcohol in Alabama Race."

112. "Phunware Announces Strategic Relationship with American Made Media Consultants for the Trump-Pence 2020 Reelection Campaign's Mobile Application Portfolio," Phunware, May 27, 2020, https://web.archive.org/web/20200926083742/https://www.phunware.com/press-releases/american-made-consultants-trump-pence-2020-relelection-campaign-mobile-application-portfolio/.

113. Sue Halpern, "How the Trump Campaign's Mobile App Is Collecting Huge Amounts of Voter Data," *The New Yorker*, September 13, 2020, https://www.newyorker.com/news/campaign-chronicles/the-trump-campaigns-mobile-app-is-collecting-massive-amounts-of-voter-data.

114. "Phunware Data Licensing," Phunware, accessed September 19, 2020, https://web.archive.org/web/20200919010301/https://www.phunware.com/data/data-licensing/.

115. Tate Ryan-Mosley, "The Technology That Powers the 2020 Campaigns, Explained," *MIT Technology Review*, September 28, 2020, https://www.technologyreview.com/2020/09/28/1008994/the-technology-that-powers-political-campaigns-in-2020-explained/; Marianna Sotomayor, "Biden Campaign's Microtargeting of Latino Communities Takes on a New Twist," *NBC News*, October 13, 2020, https://www.nbcnews.com/politics/2020-election/biden-campaign-s-microtargeting-latino-communities-takes-new-twist-n1243170.

116. Ben Popken, "Trolls for Hire: Russia's Freelance Disinformation Firms Offer Propaganda with a Professional Touch," *NBC News*, October 1, 2019, https://www.nbcnews.com/tech/security/trolls-hire-russia-s-freelance-disinformation-firms-offer-propaganda-professional-n1060781; Insikt Group, "The Price of Influence:

Disinformation in the Private Sector," Recorded Future, September 30, 2019, https://www.recordedfuture.com/disinformation-service-campaigns/.

117. Craig Silverman, Jane Lytvynenko, and William Kung, "Disinformation For Hire: How A New Breed Of PR Firms Is Selling Lies Online," *BuzzFeed News*, January 6, 2020, https://www.buzzfeednews.com/article/craigsilverman/disinformation-for-hire-black-pr-firms.

118. Samantha Bradshaw, Hannah Bailey, and Philip N. Howard, *Industrialized Disinformation: 2020 Global Inventory of Organized Social Media Manipulation* (Oxford: Computational Propaganda Research Project, January 13, 2021), https://comprop.oii.ox.ac.uk/research/posts/industrialized-disinformation/#continue.

119. Jonathan Auerbach, *Weapons of Democracy: Propaganda, Progressivism, and American Public Opinion* (Baltimore: Johns Hopkins University Press, 2015), 27.

120. Thomas Rid, *Active Measures: The Secret History of Disinformation and Political Warfare* (New York: Farrar, Straus and Giroux, 2020); Roy Godson and Richard Shultz, "Soviet Active Measures: Distinctions and Definitions," *Defense Analysis* 1, no. 2 (June 1, 1985): 101–110, https://doi.org/10.1080/07430178508405191.

121. Kaeten Mistry, "The Case for Political Warfare: Strategy, Organization and US Involvement in the 1948 Italian Election," *Cold War History* 6, no. 3 (2006): 301–329, https://doi.org/10.1080/14682740600795451; David Shimer, *Rigged: America, Russia, and One Hundred Years of Covert Electoral Interference* (New York: Knopf, 2020); Nick Fielding and Ian Cobain, "Revealed: US Spy Operation That Manipulates Social Media," *Guardian*, March 17, 2011; Stephen C. Webster, "Revealed: Air Force Ordered Software to Manage Army of Fake Virtual People," *Raw Story*, February 18, 2011; Sean Lawson, "HBGary Hearts Apple," *Forbes.com*, February 22, 2011; Nafeez Ahmed, "Your Government Wants to Militarize Social Media to Influence Your Beliefs," *Vice*, November 14, 2016, https://www.vice.com/en/article/9a384v/your-government-wants-to-militarize-social-media-to-influence-your-beliefs.

122. Lawson, "HBGary Hearts Apple."

123. Fielding and Cobain, "Revealed: US Spy Operation That Manipulates Social Media"; Webster, "Revealed: Air Force Ordered Software to Manage Army of Fake Virtual People"; Lawson, "HBGary Hearts Apple"; Ahmed, "Your Government Wants to Militarize Social Media to Influence Your Beliefs."

Chapter 8

1. Max Fisher, "Russian Hackers Find Ready Bullhorns in the Media," *New York Times*, January 8, 2017, https://www.nytimes.com/2017/01/08/world/europe/russian-hackers-find-ready-bullhorns-in-the-media.html.

2. Emma Briant, David Karpf, and Aram Sinnreich, "Beyond Cambridge Analytica: Microtargeting and Online Campaigns in 2020," Zoom panel discussion, September 2, 2020, https://www.eventbrite.com/x/117304758691/. Specifically, Briant argued that Cambridge Analytica–style targeting, particularly using fear appeals, is increasingly effective in shaping individual political opinions. Karpf was more concerned about regulatory problems and outright voter suppression. When it came to political messaging, Karpf argued that mass messaging is more effective.

3. Megan French and Natalya N. Bazarova, "Is Anybody out There?: Understanding Masspersonal Communication through Expectations for Response across Social Media Platforms," *Journal of Computer-Mediated Communication* 22, no. 6 (2017): 303–319.

4. Patrick B. O'Sullivan and Caleb T. Carr, "Masspersonal Communication: A Model Bridging the Mass-Interpersonal Divide," *New Media & Society* 20, no. 3 (March 1, 2018): 1161–1180, https://doi.org/10.1177/1461444816686104.

5. Kathleen Hall Jamieson, *Cyberwar: How Russian Hackers and Trolls Helped Elect a President: What We Don't, Can't, and Do Know* (New York: Oxford University Press, 2018), chap. 2.

6. "Revealed: Trump Campaign Strategy to Deter Millions of Black Americans from Voting in 2016," *Channel 4 News*, 2020, video, 21:39, https://www.youtube.com/watch?v=KIf5ELaOjOk.

7. "Revealed."

8. Dave Karpf, "Will the Real Psychometric Targeters Please Stand Up?" Civic Hall, February 1, 2017, https://civichall.org/civicist/will-the-real-psychometric-targeters-please-stand-up/.

9. Yochai Benkler, "The Russians Didn't Swing the 2016 Election to Trump. But Fox News Might Have," *Washington Post*, October 24, 2018, https://www.washingtonpost.com/outlook/2018/10/24/russians-didnt-swing-election-trump-fox-news-might-have/.

10. Thomas Rid, *Active Measures: The Secret History of Disinformation and Political Warfare* (New York: Farrar, Straus and Giroux, 2020), 406, 409.

11. Verizon, *Data Breach Investigations Report*, 2019.

12. "How a Hacker Convinced Motorola to Send Him Source Code," VICE: Motherboard, November 14, 2018, video, 4:15, https://www.youtube.com/watch?v=UBaVek2oTtc.

13. Benkler, Faris, and Roberts, *Network Propaganda*.

14. Miles Parks, "Florida Governor Says Russian Hackers Breached 2 Counties in 2016," NPR.org, May 14, 2019, https://www.npr.org/2019/05/14/723215498/florida

-governor-says-russian-hackers-breached-two-florida-counties-in-2016. We should note that the hack of voting machines in Florida was aided by a social engineering attack, specifically a spear-phishing attack, with the pretext of emailing from a legitimate third-party vendor.

15. David A. Mindell, "'The Clangor of That Blacksmith's Fray': Technology, War, and Experience Aboard the USS Monitor," *Technology and Culture* 36, no. 2 (1995): 242–270, https://doi.org/10.2307/3106372.

16. Peter Louis Galison, "The Ontology of the Enemy: Norbert Wiener and the Cybernetic Vision," *Critical Inquiry* 21, no 1. (autumn 1994): 228–266.

17. Thomas P. Hughes, *Rescuing Prometheus: Four Monumental Projects That Changed the Modern World* (New York: Pantheon Books, 1998).

18. John T. Correll, "Igloo White," *Air Force Magazine*, November 1, 2004; Robert R. Tomes, *U.S. Defense Strategy from Vietnam to Operation Iraqi Freedom: Military Innovation and the New American Way of War, 1973–2003* (London: Routledge, 2006).

19. Molly R. Sauter, "Kevin Mitnick, the *New York Times*, and the Media's Conception of the Hacker," in *Making Our World: The Hacker and Maker Movements in Context*, eds. Jeremy Hunsinger and Andrew Schrock (New York: Peter Lang, 2019), 21–35.

20. Aelon Porat, "Power to the People," presented at HOPE, June 26, 2020, https://infocondb.org/con/hope/hope-2020/power-to-the-people-effective-advocacy-for-privacy-and-security.

21. Gary Pruitt, "Putin Says Russia Didn't Meddle in US Vote, despite Evidence," *AP News*, June 6, 2019. https://apnews.com/article/1f12eaf734014da6bcd0995ae89c6636.

22. Krishnadev Calamur, "Putin Says 'Patriotic Hackers' May Have Targeted U.S. Election," *Atlantic*, June 1, 2017, https://www.theatlantic.com/news/archive/2017/06/putin-russia-us-election/528825/.

23. See the 25-minute mark of Alexander Nix, "From Mad Men to Math Men," presented at OMR Festival 2017, Hamburg, March 2017, video, 30:17, https://www.youtube.com/watch?v=6bG5ps5KdDo.

24. Jacob E. Osterhout, "Stealth Marketing: When You're Being Pitched and You Don't Even Know It!" nydailynews.com, April 18, 2010, https://www.nydailynews.com/life-style/stealth-marketing-pitched-don-article-1.165278.

25. Jenna Wortham, "The Well-Followed on Social Media Cash In on Their Influence," *New York Times*, June 8, 2014, https://www.nytimes.com/2014/06/09/technology/stars-of-vine-and-instagram-get-advertising-deals.html.

26. Scott Shane and Alan Blinder, "Secret Experiment in Alabama Senate Race Imitated Russian Tactics," *New York Times*, December 20, 2018, https://www.nytimes.com/2018/12/19/us/alabama-senate-roy-jones-russia.html.

27. For example, listen to the interview with Jek, a penetration tester who discusses the firing of an employee whom she social engineered. Ben Makuch and "Jek," "The Penetration Tester," October 15, 2020, in CYBER (podcast), MP3 audio, 29:00, https://podcasts.apple.com/us/podcast/re-run-the-penetration-tester/id1441708044?i=1000494881794."

28. Mark Andrejevic, *Infoglut: How Too Much Information Is Changing the Way We Think and Know* (New York: Routledge, 2013), 36.

29. Michael Bazzell and Justin Carroll, *The Complete Privacy and Security*, vol. 1. S.l, self-published (CreateSpace Independent Publishing Platform, 2016).

30. Andrejevic, *Infoglut*, 37.

31. See the 38-minute mark of Molly Schweickert, "Cambridge Analytica explains how the Trump campaign worked," *D3con: The Future of Digital Advertising*, 2017, video, 40:02, https://www.youtube.com/watch?v=bB2BJjMNXpA.

32. Sue Halpern, "How the Trump Campaign's Mobile App Is Collecting Huge Amounts of Voter Data," *New Yorker*, September 13, 2020, https://www.newyorker.com/news/campaign-chronicles/the-trump-campaigns-mobile-app-is-collecting-massive-amounts-of-voter-data.

33. "Cyberspace Solarium Commission Final Report," Washington, DC: United States Congress, March 2020, 17, https://drive.google.com/file/d/1ryMCIL_dZ30QyjFqFkkf1OMxIXJGT4yv/view?usp=embed_facebook.

34. "Cyberspace Solarium Commission Final Report," 93.

35. "Cyberspace Solarium Commission Final Report," 93.

36. Unfortunately, the Cyberspace Solarium report does not explicitly discuss opt-in regulations.

37. Charlotte Trueman, "What Impact Are Data Centres Having on Climate Change?" Computerworld, August 9, 2019, https://www.computerworld.com/article/3431148/why-data-centres-are-the-new-frontier-in-the-fight-against-climate-change.html.

38. Mél Hogan, "Data Flows and Water Woes: The Utah Data Center," *Big Data & Society* 2, no. 2 (July 2015): 8, https://doi.org/10.1177/2053951715592429.

39. Ciara Torres-Spelliscy, "Dark Money as a Political Sovereignty Problem," *King's Law Journal* 28, no. 2 (May 4, 2017): 240, https://doi.org/10.1080/09615768.2017.1351659.

40. Scott M. Cutlip, *The Unseen Power: Public Relations, A History* (Hillsdale, NJ: Lawrence Erlbaum, 1994); Torres-Spelliscy, "Dark Money as a Political Sovereignty Problem."

41. Heather K. Gerken, "Boden Lecture: The Real Problem with Citizens United: Campaign Finance, Dark Money, and Shadow Parties," *Marquette Law Review* 97, no. 4 (2014): 903–923.

42. Sarah E. Bolden, Brian McKernan, and Jennifer Stromer-Galley, "Illuminating," in "Facebook Political Advertising Transparency Report" (Syracuse University: October 5, 2020), https://news.illuminating.ischool.syr.edu/2020/10/06/facebook-political-advertising-transparency-report/.

43. Michael J. Palenchar and Kathy R. Fitzpatrick, "Secret Persuaders: Ethical and Rhetorical Perspectives on the Use of Public Relations Front Groups," in *Rhetorical and Critical Approaches to Public Relations II*, eds. Robert L. Heath, Elizabeth L. Toth, and Damion Waymer, (New York: Routledge, 2009), 280.

44. With the Supreme Court having ruled that money is speech, however, we understand that this is not likely to occur any time soon.

45. Chandra Mukerji, "Bullshitting: Road Lore Among Hitchhikers," *Social Problems* 25, no. 3 (February 1, 1978): 241–252, https://doi.org/10.2307/800062.

46. The Casebook is available at https://mediamanipulation.org/.

47. Rachelle Hampton, "The Black Feminists Who Saw the Alt-Right Threat Coming," *Slate Magazine*, April 23, 2019, https://slate.com/technology/2019/04/black-feminists-alt-right-twitter-gamergate.html.

48. Justin Lewis and Sut Jhally, "The Struggle over Media Literacy," *Journal of Communication* 48, no. 1 (1998): 109–120, https://doi.org/10.1111/j.1460-2466.1998.tb02741.x.

49. Lewis and Jhally, 110.

50. Monica Bulger and Patrick Davison, "The Promises, Challenges, and Futures of Media Literacy," *Journal of Media Literacy Education* 10, no. 1 (2018): 17.

51. Robert W. Gehl, "The Case for Alternative Social Media," *Social Media + Society* 1, no. 2 (July 1, 2015), https://doi.org/10.1177/2056305115604338.

52. Alicia Wanless and Michael Berk, "Participatory Propaganda: The Engagement of Audiences in the Spread of Persuasive Communications," presented at Social Media & Social Order, Culture Conflict 2.0, Oslo, November 2017, https://www.researchgate.net/profile/Alicia_Wanless/publication/329281610_Participatory_Propaganda_The_Engagement_of_Audiences_in_the_Spread_of_Persuasive_Communications/links/5c006978299bf1a3c1561474/Participatory-Propaganda-The-Engagement-of-Audiences-in-the-Spread-of-Persuasive-Communications.pdf.

53. Sean T. Lawson, *Cybersecurity Discourse in the United States: Cyber-Doom Rhetoric and Beyond* (London: Routledge, 2020); Bruce Schneier, "The US Has Suffered a Massive Cyberbreach. It's Hard to Overstate How Bad It Is," *Guardian*, December 23, 2020, http://www.theguardian.com/commentisfree/2020/dec/23/cyber-attack-us-security-protocols.

54. "Cyberspace Solarium Commission Final Report."

55. Christopher Wylie, *Mindf*ck: Cambridge Analytica and the Plot to Break America* (New York: Random House, 2019), 65. Note that this sounds remarkably like contemporary economic theory, which sees an economy as an emergent property of aggregated individuals' utility-maximising demand curves.

56. Christopher Simpson, *Science of Coercion: Communication Research and Psychological Warfare, 1945–1960* (New York: Oxford University Press, 1994).

57. Ross Jeffries, *Secrets Of Speed Seduction—How To Create An Instantaneous Sexual Attraction In Any Woman You Meet* (Culver City, CA: Ross Jeffries, 1994), 9.

58. "Seven Commandments of Fake News—*New York Times* Exposes Kremlin's Methods," EUvsDisinfo, November 21, 2018, https://euvsdisinfo.eu/seven-commandments-of-fake-news-new-york-times-exposes-kremlins-methods/; Bruce Schneier, "Toward an Information Operations Kill Chain," Lawfare, April 24, 2019, https://www.lawfareblog.com/toward-information-operations-kill-chain; Clint Watts, "Advanced Persistent Manipulators, Part Three: Social Media Kill Chain," *Alliance for Securing Democracy* (blog), July 22, 2019, https://securingdemocracy.gmfus.org/advanced-persistent-manipulators-part-three-social-media-kill-chain/; Rod J. Rosenstein, "Report of the Attorney General's Cyber Digital Task Force," US Department of Justice, 2018, https://www.justice.gov/archives/ag/page/file/1076696/download.

59. Sean Lawson, "Putting the 'War' in Cyberwar: Metaphor, Analogy, and Cybersecurity Discourse in the United States," *First Monday* 17, no. 7 (2012), https://doi.org/10.5210/fm.v17i7.3848; Lawson, *Cybersecurity Discourse in the United States*.

60. James W. Carey, *Communication as Culture: Essays on Media and Society* (New York: Routledge, 1989).

61. Raymond Williams, "Advertising: The Magic System," *Advertising & Society Review* 1, no. 1 (January 1, 2000), https://doi.org/10.1353/asr.2000.0016.

62. Guobin Yang, "Communication as Translation: Notes toward a New Conceptualization of Communication," in *Rethinking Media Research for Changing Societies*, eds. Matthew Powers and Adrienne Russell (New York: Cambridge University Press, 2020), 190.

63. Patricia Hill Collins, *Black Feminist Thought: Knowledge, Consciousness, and the Politics of Empowerment*, rev. 10th anniversary ed. (New York: Routledge, 2000), 30–39.

64. Collins, *Black Feminist Thought*, 38.

65. Paul Apostolidis, "Negative Dialectics and Inclusive Communication," in *Feminist Interpretations of Theodor Adorno*, ed. Renée Heberle (University Park, PA.: Penn State University Press, 2006), 245, 234.

66. Angela McRobbie, *The Uses of Cultural Studies: A Textbook* (London: SAGE, 2005), 30.

67. Collins, *Black Feminist Thought*, 36.

68. McRobbie, *The Uses of Cultural Studies*, 3.

69. See the discussions of "data for good" in Brittany Kaiser, *Targeted: The Cambridge Analytica Whistleblower's Inside Story of How Big Data, Trump, and Facebook Broke Democracy and How It Can Happen Again* (New York: HarperCollins, 2019); Wylie, *Mindf*ck*.

70. Kaiser, *Targeted*.

71. Halpern, "How the Trump Campaign's Mobile App Is Collecting Huge Amounts of Voter Data."

72. Craig Hayden and Emily T. Metzgar, *Strategic Communication and Security* (Routledge Handbooks Online, 2019), https://doi.org/10.4324/9781351180962-12; Lee Edwards, "Organised Lying and Professional Legitimacy: Public Relations' Accountability in the Disinformation Debate," *European Journal of Communication*, December 16, 2020, https://doi.org/10.1177/0267323120966851; Srividya Ramasubramanian and Omotayo O. Banjo, "Critical Media Effects Framework: Bridging Critical Cultural Communication and Media Effects through Power, Intersectionality, Context, and Agency," *Journal of Communication* 70, no. 3 (June 1, 2020): 379–400, https://doi.org/10.1093/joc/jqaa014.

Bibliography

"2-Day Social Engineering Training Course Outline." Mitnick Security Consulting, 2005. https://web.archive.org/web/20070307020902/http://www.mitnicksecurity.com/media/msc_course_outline.pdf (link no longer exists).

"III: 'Beware the Common Drinking Cup'—Reform and the Assault on the Common Drinking Cup." Disposable America (blog), December 8, 2015. https://disposableamerica.org/course-projects/a-wholesome-drink/section-iii-beware-the-common-drinking-cup-progressive-reform-and-the-assault-on-the-common-drinking-cup/.

"VI: 'No Soda Fountain Needs to Be a "Germ Exchange"'—Soda Fountains, Restaurants, and Fast Food." Disposable America (blog), December 8, 2015. https://disposableamerica.org/course-projects/a-wholesome-drink/section-vi-no-soda-fountain-needs-to-be-a-germ-exchange-soda-fountains-restaurants-and-fast-food/.

A. Ben Dump. "Computing for the Masses: A Devious Approach." *TAP*, February 1980.

Abu Arqoub, Omar Ahmad, Bahire Efe Özad, and Adeola Abdulateef Elega. "The Engineering of Consent: A State-of-the-Art Review." *Public Relations Review* 45, no. 5 (December 1, 2019). https://doi.org/10.1016/j.pubrev.2019.101830.

Ahmed, Nafeez. "Your Government Wants to Militarize Social Media to Influence Your Beliefs." *Vice*, November 14, 2016. https://www.vice.com

/en/article/9a384v/your-government-wants-to-militarize-social-media-to-influence-your-beliefs.

"Alert Number ME-000138-TT: Indicators of Compromise Pertaining to Iranian Interference in the 2020 US Presidential Election." FBI Flash, October 29, 2020. https://www.ic3.gov/Media/News/2020/201030.pdf.

Alexander, Jon, and Joachim K. H. W. Schmidt. "Social Engineering: Genealogy of a Concept." In *Social Engineering*, edited by Adam Podgórecki, Jon Alexander, and Rob Shields, 1–19. Ottawa: Carleton University Press, 1996.

Allison, Bill, and Misyrlena Egkolfopoulou. "Trump Outpaces Biden in Zeroing In on Voters With Facebook Tools." Bloomberg.com, July 13, 2020. https://www.bloomberg.com/news/articles/2020-07-13/trump-more-than-biden-is-tapping-into-facebook-targeting-tools.

Almog, Ran, and Danny Kaplan. "The Nerd and His Discontent: The Seduction Community and the Logic of the Game as a Geeky Solution to the Challenges of Young Masculinity." *Men and Masculinities* 20, no. 1 (April 1, 2017): 27–48. https://doi.org/10.1177/1097184X15613831.

Amer, Karim, and Jehane Noujaim, dirs.. *The Great Hack*. Documentary. The Othrs, 2019. https://www.netflix.com/title/80117542.

Andrejevic, Mark. *Infoglut: How Too Much Information Is Changing the Way We Think and Know*. New York: Routledge, 2013.

Andreoletti, Davide, and Enrico Frumento. "Social Engineering to the Extreme: The Cambridge Analytica Case." Dogana Project, March 23, 2018. https://www.dogana-project.eu/index.php/social-engineering-blog/11-social-engineering/92-cambridge-analytica.

anonymous. "'Index of' inurl:'/$Recycle.Bin/'." Exploit Database, May 5, 2017. https://www.exploit-db.com/ghdb/4463/.

anonymous. "'index of' inurl:recycler." Exploit Database, May 4, 2004. https://www.exploit-db.com/ghdb/205/.

anonymous. "inurl:trash intitle:index.of." Exploit Database, June 6, 2016. https://www.exploit-db.com/ghdb/4295/.

Apostolidis, Paul. "Negative Dialectics and Inclusive Communication." In *Feminist Interpretations of Theodor Adorno*, edited by Renée Heberle, 233–256. University Park, PA.: Penn State University Press, 2006.

Ashcraft, Karen Lee, and Brenda J. Allen. "The Racial Foundation of Organizational Communication." *Communication Theory* 13, no. 1 (2003): 5–38. https://doi.org/10.1111/j.1468-2885.2003.tb00280.x.

Ashcraft, Karen Lee, Timothy R. Kuhn, and François Cooren. "Constitutional Amendments: 'Materializing' Organizational Communication." *The Academy of Management Annals* 3, no. 1 (2009): 1–64. https://doi.org/10.1080/19416520903047186.

"Assessing Russian Activities and Intentions in Recent US Elections." Washington, DC: Office of the Director of National Intelligence, 2017. https://www.dni.gov/files/documents/ICA_2017_01.pdf.

Auerbach, Jonathan. *Weapons of Democracy: Propaganda, Progressivism, and American Public Opinion*. Baltimore: Johns Hopkins University Press, 2015.

Autovon, David. "The Automatic Wiretap: Are Your Telephone Conversations Really Private?" *Telephone Electronics Line*, December 1974.

Baker, Ray Stannard. "Railroads on Trial." *McClure's Magazine*, March 1906.

Bandler, Richard, John Grinder, and Steve Andreas. *Frogs into Princes: Neuro Linguistic Programming*. Moab, UT: Real People Press, 1979.

Bandler, Richard, John Grinder, Steve Andreas, and Connirae Andreas. *Reframing: Neuro-Linguistic Programming [Trade Mark Symbol] and the Transformation of Meaning*. Moab, UT: Real People Press, 1982.

Barry, Ellen. "Long Before Cambridge Analytica, a Belief in the 'Power of the Subliminal' (Published 2018)." *New York Times*, April 20, 2018. https://www.nytimes.com/2018/04/20/world/europe/oakes-scl-cambridge-analytica-trump.html.

Baym, Nancy K., Yan Bing Zhang, and Mei-Chen Lin. "Social Interactions Across Media: Interpersonal Communication on the Internet, Telephone and Face-to-Face." *New Media & Society* 6, no. 3 (June 1, 2004): 299–318. https://doi.org/10.1177/1461444804041438.

Bazzell, Michael. "142-OSINT Extravaganza and Book Release!" 2019. In The Privacy, Security, and OSINT Show. MP3 audio, 30:30. https://soundcloud.com/user-98066669/142-osint-extravaganza-and-book-release.

Bazzell, Michael. *Open Source Intelligence Techniques: Resources for Searching and Analyzing Online Information*. 7th ed. Self-published, CreateSpace Independent Publishing Platform, 2019.

Bazzell, Michael and Justin Carroll. *The Complete Privacy and Security Desk Reference*. Vol. 1. S.l. Self-published, CreateSpace Independent Publishing Platform, 2016.

Benkler, Yochai. "The Russians Didn't Swing the 2016 Election to Trump. But Fox News Might Have." *Washington Post*, October 24, 2018. https://www.washingtonpost.com/outlook/2018/10/24/russians-didnt-swing-election-trump-fox-news-might-have/.

Benkler, Yochai, Robert Faris, and Hal Roberts. *Network Propaganda: Manipulation, Disinformation, and Radicalization in American Politics*. New York: Oxford University Press, 2018.

Berger, Charles R. "Interpersonal Communication." In *The International Encyclopedia of Communication*. International Communication Association, 2010. https://doi.org/10.1002/9781405186407.wbieci077.

Bernays, Edward L. *Biography of an Idea: Memoirs of Public Relations Counsel*. New York: Simon & Schuster, 1965. http://archive.org/details/biographyofideam00bern.

Bernays, Edward L. *Crystallizing Public Opinion*. 1923.

Bernays, Edward L. *The Engineering of Consent*. Philadelphia, 1947.

Bernays, Edward L. "The Engineering of Consent." *The ANNALS of the American Academy of Political and Social Science* 250, no. 1 (March 1947): 113–120. https://doi.org/10.1177/000271624725000116.

Bernays, Edward L. "Human Engineering and Social Adjustment." *ETC: A Review of General Semantics* 74, no. 3–4 (July 2017): 346–351.

Bernays, Edward L. "The Marketing of National Policies: A Study of War Propaganda." *Journal of Marketing* 6, no. 3 (1942): 236–244. https://doi.org/10.1177/002224294200600303.

Bernays, Edward L. *Propaganda*. New York: H. Liveright, 1928.

Bernays, Edward L. "Torches-031710." SpectorPR, 2010. Video, 6:29. https://www.youtube.com/watch?time_continue=299&v=6pyyP2chM8k&feature=emb_logo.

Bernays, Edward L., and Howard Walden Cutler, eds. *The Engineering of Consent*. Norman: University of Oklahoma Press, 1955.

Bernays, Edward L. "Theory and Practice of Public Relations: A Resume." In *The Engineering of Consent*, edited by Edward L. Bernays and Howard Walden Cutler, 3–25. Norman: University of Oklahoma Press, 1955.

Bernays, Edward, and Mark Crispin Miller. *Propaganda*. First paperback ed. Brooklyn: Ig Publishing, 2004.

Beskow, David M., and Kathleen M. Carley. "Social Cybersecurity: An Emerging National Security Requirement." *Military Review: The Professional Journal of the U.S. Army*, April 2019.

Bilton, Alan. *Silent Film Comedy and American Culture*. New York: Springer, 2013.

Bineham, Jeffery L. "A Historical Account of the Hypodermic Model in Mass Communication." *Communication Monographs* 55, no. 3 (September 1988): 230–246. https://doi.org/10.1080/03637758809376169.

BIOC Agent 003. "Course in Basic Telecommunications Part I," November 15, 1983. http://textfiles.com/phreak/BIOCAGENT/bioc1.wri.

BIOC Agent 003. "Course in Basic Telecommunications Part II." Archive. Textfiles.com, July 18, 1984. https://textfiles.com/phreak/BIOCAGENT/basicom2.phk.

BIOC Agent 003. "Course in Basic Telecommunications Part II," December 8, 1983. https://textfiles.com/phreak/BIOCAGENT/bioc2.wri.

BIOC Agent 003. "Course in Basic Telecommunications Part IV," April 13, 1984. https://textfiles.com/phreak/BIOCAGENT/bioc2.wri.

BIOC Agent 003 and Tharrys Ridenow. "Word-Processed Redoing of Bioc Agent 003's Course in Telecommunications," June 18, 1984. http://textfiles.com/phreak/BIOCAGENT/bioc003.001.

Bivins, Thomas H. "A Golden Opportunity? Edward Bernays and the Dilemma of Ethics." *American Journalism* 30, no. 4 (January 1, 2013): 496–519. https://doi.org/10.1080/08821127.2013.857981.

Bolden, Sarah E., Brian McKernan, and Jennifer Stromer-Galley. "Illuminating." In "Facebook Political Advertising Transparency Report." Syracuse University: October 5, 2020. https://news.illuminating.ischool.syr.edu/2020/10/06/facebook-political-advertising-transparency-report/.

Bond, Shannon. "Facebook And Twitter Limit Sharing 'New York Post' Story About Joe Biden." *NPR.org*, October 14, 2020. https://www.npr

.org/2020/10/14/923766097/facebook-and-twitter-limit-sharing-new-york-post-story-about-joe-biden.

Bos, Martin, David Kennedy, Justin Elze, and Scott White. "Ashley Madison Hacked. Dump Released." *TrustedSec* (blog), August 19, 2015. https://www.trustedsec.com/blog/ashley-madison-database-dumped/.

Bossetta, Michael. "The Weaponization of Social Media: Spear Phishing and Cyberattacks on Democracy." *Journal of International Affairs* 70, no. 1.5 (2018): 97–106.

Bradshaw, Samantha, Hannah Bailey, and Philip N. Howard. *Industrialized Disinformation: 2020 Global Inventory of Organized Social Media Manipulation*. Oxford: Computational Propaganda Research Project, January 13, 2021. https://comprop.oii.ox.ac.uk/research/posts/industrialized-disinformation/#continue.

Braverman, Harry. *Labor and Monopoly Capital; the Degradation of Work in the Twentieth Century*. New York: Monthly Review Press, 1974.

Brennan, Geoffrey, and James M. Buchanan. *The Reason of Rules: Constitutional Political Economy*. Vol. 10 of *The Collected Works of James M. Buchanan*. Indianapolis: Liberty Fund, 2000.

Briant, Emma, David Karpf, and Aram Sinnreich. "Beyond Cambridge Analytica: Microtargeting and Online Campaigns in 2020." Zoom panel discussion, September 2, 2020. https://www.eventbrite.com/x/117304758691/.

Bristow, Nancy K. *Making Men Moral: Social Engineering during the Great War*. New York: New York University Press, 1996.

Brown, Steven D. "In Praise of the Parasite: The Dark Organizational Theory of Michel Serres." *Informática Na Educação: Teoria & Prática* 16, no. 1 (2013).

Brown, Steven D. "Michel Serres: Science, Translation and the Logic of the Parasite." *Theory, Culture & Society* 19, no. 3 (2002): 1–27. https://doi.org/10.1177/026327602401081503.

Bruce, Kyle. "Henry S. Dennison, Elton Mayo, and Human Relations Historiography." *Management & Organizational History* 1, no. 2 (May 1, 2006): 177–199. https://doi.org/10.1177/1744935906064095.

Buckley Jr., William F. "Our Mission Statement." *National Review*, November 19, 1955. https://www.nationalreview.com/1955/11/our-mission-statement-william-f-buckley-jr/.

Bulger, Monica, and Patrick Davison. "The Promises, Challenges, and Futures of Media Literacy." *Journal of Media Literacy Education* 10, no. 1 (2018): 1–21.

Burke, Garance. "Financially Troubled Startup Helped Power Trump Campaign." *AP News*, November 17, 2020. https://apnews.com/article/phunware-app-helped-power-trump-campaign-89ed273f60e37ff9ee020dd2f5d3df04.

Butsch, Richard. *The Citizen Audience: Crowds, Publics, and Individuals.* London: Routledge, 2007.

Cadwalladr, Carole. "'I Made Steve Bannon's Psychological Warfare Tool': Meet the Data War Whistleblower." *Guardian*, March 18, 2018. https://www.theguardian.com/news/2018/mar/17/data-war-whistleblower-christopher-wylie-faceook-nix-bannon-trump.

Cadwalladr, Carole, and Emma Graham-Harrison. "How Cambridge Analytica Turned Facebook 'Likes' into a Lucrative Political Tool." *Guardian*, March 17, 2018. https://www.theguardian.com/technology/2018/mar/17/facebook-cambridge-analytica-kogan-data-algorithm.

Cadwalladr, Carole, and Emma Graham-Harrison. "Revealed: 50 Million Facebook Profiles Harvested for Cambridge Analytica in Major Data Breach." *Guardian*, March 17, 2018. https://www.theguardian.com/news/2018/mar/17/cambridge-analytica-facebook-influence-us-election.

Cadwalladr, Carole, and Mark Townsend. "Revealed: The Ties That Bound Vote Leave's Data Firm to Controversial Cambridge Analytica." *Guardian*, March 24, 2018. https://www.theguardian.com/uk-news/2018/mar/24/aggregateiq-data-firm-link-raises-leave-group-questions.

Calamur, Krishnadev. "Putin Says 'Patriotic Hackers' May Have Targeted U.S. Election." *The Atlantic*, June 1, 2017. https://www.theatlantic.com/news/archive/2017/06/putin-russia-us-election/528825/.

"Cambridge Analytica Uncovered: Secret Filming Reveals Election Tricks." *Channel 4 News*, 2018. Video, 19:12. https://www.youtube.com/watch?v=mpbeOCKZFfQ.

Cameron, Dell. "It's Scary How Much Personal Data People Leave on Used Laptops and Phones, Researcher Finds." Gizmodo, March 19, 2019. https://gizmodo.com/its-scary-how-much-personal-data-people-leave-on-used-l-1833383903.

Campbell, John Martin. *Slinging the Bull in Korea: An Adventure in Psychological Warfare*. Albuquerque: University of New Mexico Press, 2010.

Carey, James W. *Communication as Culture: Essays on Media and Society*. New York: Routledge, 1989.

Carlson, Robert O. "The Use of Public Relations Research by Large Corporations." *Public Opinion Quarterly* 21, no. 3 (1957): 341. https://doi.org/10.1086/266726.

Carnegie, Dale. *How to Win Friends & Influence People*. 80th anniversary ed. New York: Simon & Schuster, 2017.

CBS This Morning. "Former Cambridge Analytica Employee Says 'There Were More' Apps That Collected User Data." CBS, June 27, 2018. https://www.cbsnews.com/news/cambridge-analytica-brittany-kaiser-speaks-out-data-collection-facebook/.

Chase, W. Howard. "Nothing Just Happens, Somebody Makes It Happen." *Public Relations Journal* (November 1962): 5–11.

Cialdini, Robert B. *Influence: The Psychology of Persuasion*. New York: Quill, 1985.

Clark, A. B. "The Development of Telephony in the United States." *Transactions of the American Institute of Electrical Engineers, Part I: Communication and Electronics* 71, no. 5 (November 1952): 348–364. https://doi.org/10.1109/TCE.1952.6371872.

Clark, Victoria, Mikhaila Fogel, Susan Hennessey, Quinta Jurecic, Matthew Kahn, and Benjamin Wittes. "Russian Electoral Interference: 2018 Midterms Edition." *Lawfare* (blog), October 19, 2018. https://www.lawfareblog.com/russian-electoral-interference-2018-midterms-edition.

"Clinton Foundation Archives." FactCheck.org, 2020. https://www.factcheck.org/issue/clinton-foundation/.

Cloud, Dana L. *Control and Consolation in American Culture and Politics: Rhetorics of Therapy*. Thousand Oaks, CA: Sage Publications, 1998.

Cohen, G. A. "Deeper into Bullshit." In *The Contours of Agency: Essays on Themes from Harry Frankfurt*, edited by Sarah Buss and Lee Overton, 321–339. Cambridge, MA: MIT Press, 2002.

Coleman, E. Gabriella, and Alex Golub. "Hacker Practice: Moral Genres and the Cultural Articulation of Liberalism." *Anthropological Theory* 8, no. 3 (2008): 255–277.

Collins, Patricia Hill. *Black Feminist Thought: Knowledge, Consciousness, and the Politics of Empowerment*. Revised 10th anniversary ed. New York: Routledge, 2000.

Conheady, Sharon. "The Future of Social Engineering." Presented at DeepSec 2010, Vienna, November 26, 2010. Video, 54:16. https://www.youtube.com/watch?v=bzWRtxA5DCo.

Conheady, Sharon. "Social Engineering for Penetration Testers." Presented at BruCON Security Conference, Ghent, Belgium, October 3, 2018. Video, 1:02:23. https://www.youtube.com/watch?v=oAJ1pNJnJHQ.

Conheady, Sharon. *Social Engineering in IT Security: Tools, Tactics, and Techniques*. New York: McGraw-Hill Education, 2014.

Connell, Raewyn. "Accountable Conduct: 'Doing Gender' in Transsexual and Political Retrospect." *Gender & Society* 23, no. 1 (2009): 104–111.

Cooke, Morris L. "The Spirit and Social Significance of Scientific Management." *Journal of Political Economy* 21, no. 6 (June 1913): 481–493. https://doi.org/10.1086/252258.

Correll, John T. "Igloo White." *Air Force Magazine*, November 1, 2004.

Cullison, Alan, Rebecca Ballhaus, and Dustin Volz. "Trump Repeatedly Pressed Ukraine President to Investigate Biden's Son." *Wall Street Journal*, September 21, 2019. https://www.wsj.com/articles/trump-defends-conversation-with-ukraine-leader-11568993176.

Curious. "Dear 2600." *2600: The Hacker Quarterly*, June 14, 1984.

Cutlip, Scott M. *The Unseen Power: Public Relations, A History*. Hillsdale, NJ: Lawrence Erlbaum, 1994.

Cybersecurity & Infrastructure Security Agency. "Alert (AA20–304A): Iranian Advanced Persistent Threat Actor Identified Obtaining Voter Registration Data," October 30, 2020. https://us-cert.cisa.gov/ncas/alerts/aa20-304a.

"Cyberspace Solarium Commission Final Report." Washington, DC: United States Congress, March 2020. https://drive.google.com/file/d/1ryMCIL_dZ30QyjFqFkkf10MxIXJGT4yv/view?usp=embed_facebook.

Das, Sauvik, and Adam Kramer. "Self-Censorship on Facebook." *Proceedings of the Seventh International AAAI Conference on Weblogs and Social Media* (2013): 8.

Davies, Harry. "Ted Cruz Using Firm That Harvested Data on Millions of Unwitting Facebook Users." *Guardian*, December 11, 2015. https://www.theguardian.com/us-news/2015/dec/11/senator-ted-cruz-president-campaign-facebook-user-data.

Davis, Joseph S. "Statistics and Social Engineering." *Journal of the American Statistical Association* 32, no. 197 (1937): 1–7.

Dawson, Andrew, and Martin Innes. "How Russia's Internet Research Agency Built Its Disinformation Campaign." *The Political Quarterly* 90, no. 2 (2019): 245–256. https://doi.org/10.1111/1467-923X.12690.

DeLozier, Judith and John Grinder. *Turtles All the Way Down: Prerequisites to Personal Genius*. N.p.: Metamorphous Press, 1996.

De Payne, Lewis. "Sensual Access: The High Tech Guide to Seducing Women Using Your Home Computer," n.d. https://www.scribd.com/document/174590270/Ross-Jeffries-Sensual-Access.

Dennison, Henry S. *Organization engineering*. New York: McGraw-Hill, 1931.

DiAquoi, Raygine. "Symbols in the Strange Fruit Seeds: What 'the Talk' Black Parents Have with Their Sons Tells Us About Racism." *Harvard Educational Review* 87, no. 4 (December 1, 2017): 512–537. https://doi.org/10.17763/1943-5045-87.4.512.

DiResta, Renee, Kris Shaffer, Becky Ruppel, David Sullivan, Robert Matney, Ryan Fox, Jonathan Albright, and Ben Johnson. "The Tactics & Tropes of the Internet Research Agency." New Knowledge [now Yonder], 2018.

Douglas, Mary. *Purity and Danger: An Analysis of Concepts of Pollution and Taboo*. London: Penguin, 1966.

Dreeke, Robin K. *It's Not All about "Me": The Top Ten Techniques for Building Quick Rapport with Anyone*. Self-published, People Formula, 2011.

Duguay, Stefanie. "Dressing up Tinderella: Interrogating Authenticity Claims on the Mobile Dating App Tinder." *Information, Communication & Society* 20, no. 3 (March 4, 2017): 351–367. https://doi.org/10.1080/1369118X.2016.1168471.

Earp, Edwin Lee. *The Social Engineer*. New York, Eaton & Mains; Cincinnati, Jennings & Graham, 1911. http://archive.org/details/cu31924014043370.

"Easter Sun Finds the Past in Shadow at Modern Parade." *New York Times*. March 1, 1929.

Edwards, Lee. "Organised Lying and Professional Legitimacy: Public Relations' Accountability in the Disinformation Debate." *European Journal of Communication*, December 16, 2020. https://doi.org/10.1177/0267323120966851.

Eighner, Lars. "On Dumpster Diving." *The Threepenny Review*, no. 47 (1991): 6–8.

Ekman, Paul. *Emotions Revealed: Understanding Faces and Feelings*. London: Orion, 2004.

Ekman, Paul, and Wallace V. Friesen. *Unmasking the Face: A Guide to Recognizing Emotions from Facial Expressions*. Los Angeles: Malor, 2003.

Ewen, Stuart. *Captains of Consciousness: Advertising and the Social Roots of the Consumer Culture*. New York: McGraw-Hill, 1976.

Ewen, Stuart. *PR! A Social History of Spin*. New York: Basic Books, 1996.

Fenwick, Sheridan. *Getting It: The Psychology of Est*. Philadelphia: Lippincott, 1976. http://archive.org/details/gettingitpsychol00fenw.

Fielding, Nick, and Ian Cobain. "Revealed: US Spy Operation That Manipulates Social Media." *Guardian*, March 17, 2011.

"Find What's Changed in a File—Computer—Docs Editors Help." Google, 2020. https://support.google.com/docs/answer/190843?co=GENIE.Platform%3DDesktop&hl=en.

Fischer, Claude S. *America Calling: A Social History of the Telephone to 1940*. Berkeley: University of California Press, 1992.

Fisher, Max. "Russian Hackers Find Ready Bullhorns in the Media." *New York Times*, January 8, 2017. https://www.nytimes.com/2017/01/08/world/europe/russian-hackers-find-ready-bullhorns-in-the-media.html.

Fitterer, Jason. "Putting a Lid on Online Dumpster-Diving: Why the Fair and Accurate Credit Transactions Act Should Be Amended to Include E-Mail Receipts." *Northwestern Journal of Technology and Intellectual Property* 9, no. 8 (2011): 591–606.

Fleischman, Doris E. *An Outline of Careers for Women; a Practical Guide to Achievement*. Garden City, NY: Doubleday, Doran, 1928. https://catalog.hathitrust.org/Record/001106416.

Fleischman, Doris E. "Public Relations and the Consumer." Presented at the Fashion Group, New York, October 30, 1935.

Florman, Samuel C. *The Existential Pleasures of Engineering*. New York: St. Martin's, 1996.

Flying Penguin. "Flying Penguin Presents: Bullshitting the Operator." In *The Computer Underground: Computer Hacking, Crashing, Pirating, and Phreaking*, edited by M. Harry. Port Townsend, WA: Loompanics Unlimited, 1985.

Fogg, B. J. "Mass Interpersonal Persuasion: An Early View of a New Phenomenon." *Persuasive Technology*, 2008, 23–34.

Fonesca, Brian. "Kevin Mitnick, Hacker Extraordinaire, Speaks out on Security in Today's Internet Age." *InfoWorld*, December 1, 2000. https://www.computerworld.com/article/2784030/kevin-mitnick--the-hacker-extraordinaire-speaks-out-on-security.html.

"Foreign Threats to the 2020 US Federal Elections." Washington, DC: National Intelligence Council, March 10, 2021. https://www.dni.gov/files/ODNI/documents/assessments/ICA-declass-16MAR21.pdf.

Frankfurt, Harry G. *On Bullshit*. Princeton, NJ: Princeton University Press, 2005.

Frantz, Josh. "Exfiltrating Remaining Private Information from Donated Devices." Rapid7 (blog), March 19, 2019. https://blog.rapid7.com/2019/03/19/buy-one-device-get-data-free-private-information-remains-on-donated-devices/.

French, Megan, and Natalya N. Bazarova. "Is Anybody out There?: Understanding Masspersonal Communication through Expectations for Response across Social Media Platforms." *Journal of Computer-Mediated Communication* 22, no. 6 (2017): 303–319.

Friedman, Milton, and Rose D. Friedman. *Capitalism and Freedom*. Chicago: University of Chicago Press, 2002.

Galison, Peter Louis. "The Ontology of the Enemy: Norbert Wiener and the Cybernetic Vision." *Critical Inquiry* 21, no 1. (Autumn 1994): 228–266.

Gallagher, Erin. "Introduction: Memetic Warfare." Medium, July 29, 2018. https://medium.com/@erin_gallagher/alt-right-culture-jamming-and-memetic-warfare-93b646263f7d.

Gehl, Robert W. "The Case for Alternative Social Media." *Social Media + Society* 1, no. 2 (July 1, 2015). https://doi.org/10.1177/2056305115604338.

Gehl, Robert W., and Maria Bakardjieva. "Socialbots and Their Friends." In *Socialbots and Their Friends: Digital Media and the Automation of Sociality*, edited by Robert W. Gehl and Maria Bakardjieva, 1–16. London: Routledge, 2016.

Geltzer, Joshua. "Don't Be Fooled: There *Was* Election Interference in 2018." *Just Security* (blog), November 7, 2018. https://www.justsecurity.org/61372/dont-fooled-was-election-interference-2018/.

Gerard, Judas. "CN/a." *TAP*, October 1982. http://www.textfiles.com/tap/issue.78/tap7803.jpg.

Gerken, Heather K. "Boden Lecture: The Real Problem With Citizens United: Campaign Finance, Dark Money, and Shadow Parties." *Marquette Law Review* 97, no. 4 (2014): 903–923.

Gerstein, Josh. "U.S. Brings First Charge for Meddling in 2018 Midterm Elections." *POLITICO*, October 19, 2018. https://politi.co/2Ajbubq.

"Getting 2 Equal: United Not Divided." State of Black America. New York: National Urban League, 2019. http://soba.iamempowered.com/2019-report.

Gibbs, Jennifer L., Dina Nekrassova, Svetlana V. Grushina, and Sally Abdul Wahab. "Reconceptualizing Virtual Teaming from a Constitutive Perspective Review, Redirection, and Research Agenda." *Annals of the International Communication Association* 32, no. 1 (January 2008): 187–229. https://doi.org/10.1080/23808985.2008.11679078.

"Gin, n.1." In *OED Online*. Oxford University Press. Accessed January 7, 2019. http://www.oed.com/view/Entry/78357.

Gitelman, Lisa. *Paper Knowledge: Toward a Media History of Documents*. Durham, NC: Duke University Press, 2014.

Godson, Roy, and Richard Shultz. "Soviet Active Measures: Distinctions and Definitions." *Defense Analysis* 1, no. 2 (June 1, 1985): 101–110. https://doi.org/10.1080/07430178508405191.

Gokulan, Dhanusha. "Famous Hacker Proves Your Firm's System Can Be Accessed in an Hour." *Khaleej Times*, April 1, 2019. https://www.khaleejtimes.com/technology/famous-hacker-proves-your-firms-system-can-be-accessed-in-an-hour.

Goldman, Adam. "Justice Dept. Accuses Russians of Interfering in Midterm Elections." *New York Times*, October 19, 2018. https://www.nytimes.com/2018/10/19/us/politics/russia-interference-midterm-elections.html.

Goldman, Adam, Julian E. Barnes, Maggie Haberman, and Nicholas Fandos. "Lawmakers Are Warned That Russia Is Meddling to Re-Elect Trump." *New York Times*, February 20, 2020. https://www.nytimes.com/2020/02/20/us/politics/russian-interference-trump-democrats.html.

Goodell, Jeff. *The Cyberthief and the Samurai: The True Story of Kevin Mitnick, and the Man Who Hunted Him Down*. New York: Dell, 1996.

Goodell, Jeff. "The Samurai and the Cyberthief." *Rolling Stone*, no. 707 (May 4, 1995): 40.

Goodman, Amy, Emma Briant, Karim Amer, Brittany Kaiser, and Jehane Noujaim. "The Weaponization of Data: Cambridge Analytica, Information Warfare & the 2016 Election of Trump." *Democracy Now!*, January 10, 2020. Video, 59:02. https://www.democracynow.org/2020/1/10/defense_contractors_are_using_a_new.

Green, Michael Z. "Against Employer Dumpster-Diving for Email." *South Carolina Law Review* 64, no. 2 (2012): 323–368.

Grifter. "Dumpster Diving: One Man's Trash . . ." Archive. Textfiles.com, 2002. http://web.textfiles.com/hacking/dumpster_diving.txt.

Grut, Ståle. "OSINT Journalism Goes Mainstream." *NiemanLab* (blog), January 3, 2020. https://www.niemanlab.org/2020/01/osint-journalism-goes-mainstream/.

Gunderman, Richard. "The Manipulation of the American Mind: Edward Bernays and the Birth of Public Relations." The Conversation, July 9, 2015.

http://theconversation.com/the-manipulation-of-the-american-mind-edward-bernays-and-the-birth-of-public-relations-44393.

Hadnagy, Christopher. "Social Engineering Code of Ethics." *Security Through Education* (blog), April 4, 2019. https://www.social-engineer.org/framework/general-discussion/social-engineering-code-of-ethics/.

Hadnagy, Christopher. *Social Engineering: The Art of Human Hacking.* Indianapolis: Wiley, 2010.

Hadnagy, Christopher. *Social Engineering: The Science of Human Hacking.* 2nd ed. Indianapolis: Wiley, 2018.

Hadnagy, Christopher, and Paul Ekman. *Unmasking the Social Engineer: The Human Element of Security.* Edited by Paul F. Kelly. Indianapolis: Wiley, 2014.

Hadnagy, Chris, Michele Fincher, and Jordan Harbinger. "Ep 094—The Art of Charm Imitates Life." June 12, 2017. In Security through Education (podcast). MP3 audio, 57:10. https://www.social-engineer.org/podcast/ep-094-art-charm-imitates-life/.

Hadnagy, Chris, Robin Dreeke, and Perry Carpenter. "Ep 120—Sizing People Up—LIVE AT DEF CON 27 with Robin Dreeke." August 19, 2019. In Security through Education (podcast). MP3 audio, 01:09:14. https://www.social-engineer.org/podcast/ep-120-sizing-people-up-live-at-def-con-27-with-robin-dreeke/.

Hafner, Katie, and John Markoff. *Cyberpunk: Outlaws and Hackers on the Computer Frontier.* New York: Simon & Schuster, 1995.

Hallahan, Kirk. "Ivy Lee and the Rockefellers' Response to the 1913–1914 Colorado Coal Strike." *Journal of Public Relations Research* 14, no. 4 (October 1, 2002): 265–315. https://doi.org/10.1207/S1532754XJPRR1404_1.

Hallahan, Kirk, Derina Holtzhausen, Betteke van Ruler, Dejan Verčič, and Krishnamurthy Sriramesh. "Defining Strategic Communication." *International Journal of Strategic Communication* 1, no. 1 (March 22, 2007): 3–35. https://doi.org/10.1080/15531180701285244.

Halpern, Sue. "How the Trump Campaign's Mobile App Is Collecting Huge Amounts of Voter Data." *The New Yorker*, September 13, 2020. https://www.newyorker.com/news/campaign-chronicles/the-trump-campaigns-mobile-app-is-collecting-massive-amounts-of-voter-data.

Hampton, Rachelle. "The Black Feminists Who Saw the Alt-Right Threat Coming." *Slate Magazine*, April 23, 2019. https://slate.com/technology/2019/04/black-feminists-alt-right-twitter-gamergate.html.

Harbinger, Jordan, and Marni. "Marni | What Women Think About Confident Men (episode 94)." May 29, 2009. In Art of Charm (podcast). MP3 audio, 45:25. https://www.youtube.com/watch?v=DVXICSQtR0g.

Harbinger, Jordan, and Robert Glover. "Dr. Robert Glover | *No More Mr. Nice Guy* (episode 145)." January 16, 2012. In Art of Charm (podcast). MP3 audio, 1:06:35. https://www.youtube.com/watch?v=EJp5R_6HOA0.

Harbinger, Jordan and Susan Kuchinskas. "Susan Kuchinskas, *The Chemistry of Connection* (Episode #137)." October 3, 2011. In Art of Charm (podcast). MP3 audio, 48:16. https://www.youtube.com/watch?v=7yNhtQUk9S8.

Harford, Tim. "Big Data: A Big Mistake?" *Significance* 11, no. 5 (2014): 14–19. https://doi.org/10.1111/j.1740-9713.2014.00778.x.

Harvey, David. *A Brief History of Neoliberalism*. Oxford: Oxford University Press, 2005.

Hassard, John S. "Rethinking the Hawthorne Studies: The Western Electric Research in Its Social, Political and Historical Context." *Human Relations* 65, no. 11 (October 1, 2012): 1431–1461. https://doi.org/10.1177/0018726712452168.

Hatfield, Joseph M. "Social Engineering in Cybersecurity: The Evolution of a Concept." *Computers & Security* 73 (March 1, 2018): 102–113. https://doi.org/10.1016/j.cose.2017.10.008.

Havel, Hippolyte. "The Civil War in Colorado." *Mother Earth* 9 (March 1914): 71–77.

Hayden, Craig, and Emily T. Metzgar. *Strategic Communication and Security*. Routledge Handbooks Online, 2019. https://doi.org/10.4324/9781351180962-12.

Headley, Susan. "Social Engineering and Psychological Subversion of Trusted Systems." Presented at DEF CON, Las Vegas, August 4, 1995.

Hebbard, B., P. Grosso, T. Baldridge, C. Chan, D. Fishman, P. Goshgarian, T. Hilton, J. Hoshen, K. Hoult, G. Huntley, M. Stolarchuk, and L. Warner. "A Penetration Analysis of the Michigan Terminal System." *ACM SIGOPS*

Operating Systems Review 14, no. 1 (January 1, 1980): 7–20. https://doi.org/10.1145/850693.850694.

Hendriks, Eric C. "Ascetic Hedonism: Self and Sexual Conquest in the Seduction Community." *Cultural Analysis* 11 (2012): 1–14.

Henry, Susan. "'There Is Nothing in This Profession . . . That a Woman Cannot Do': Doris E. Fleischman and the Beginnings of Public Relations." *American Journalism* 16, no. 2 (April 1, 1999): 85–111. https://doi.org/10.1080/08821127.1999.10739176.

Hess, Michael. "Security Tips from a Legendary Hacker." CBS News, December 19, 2011. https://www.cbsnews.com/news/security-tips-from-a-legendary-hacker/.

Hetherington, Kevin, and Nick Lee. "Social Order and the Blank Figure." *Environment and Planning D: Society and Space* 18, no. 2 (2000): 169–184.

Hiebert, Ray Eldon. *Courtier to the Crowd; the Story of Ivy Lee and the Development of Public Relations*. Ames, IA: Iowa State University Press, 1966. http://archive.org/details/courtiertocrowds0000unse.

Hill, Fiona. "Opinion | The Biggest Risk to This Election Is Not Russia. It's Us." *New York Times*, October 7, 2020. https://www.nytimes.com/2020/10/07/opinion/trump-russia-election-interference.html.

Hird, Myra J. "Knowing Waste: Towards an Inhuman Epistemology." *Social Epistemology* 26, no. 3–4 (2012): 453–469. https://doi.org/10.1080/02691728.2012.727195.

Hoffman, John. *The Art & Science of Dumpster Diving*. Port Townsend, WA: Loompanics, 1993.

Hogan, Mél. "Data Flows and Water Woes: The Utah Data Center." *Big Data & Society* 2, no. 2 (July 2015). https://doi.org/10.1177/2053951715592429.

Hogan, Mél. "The Archive as Dumpster." *Pivot: A Journal of Interdisciplinary Studies & Thought* 4, no. 1 (2015). https://doi.org/10.25071/2369-7326.39565.

"How a Hacker Convinced Motorola to Send Him Source Code." VICE: Motherboard, November 14, 2018. Video, 4:15. https://www.youtube.com/watch?v=UBaVek2oTtc.

"How to S.E. Your Sexy Back." Social-Engineer.org, January 8, 2012. https://www.social-engineer.org/podcast/episode-030-how-to-s-e-your-sexy-back/.

Howard, Philip N. *Lie Machines: How to Save Democracy from Troll Armies, Deceitful Robots, Junk News Operations, and Political Operatives*. New Haven, CT: Yale University Press, 2020.

Howard, Philip N., Bharath Ganesh, Dimitra Liotsiou, John Kelly, and Camille François. "The IRA, Social Media and Political Polarization in the United States, 2012–2018." Oxford: Computational Propaganda Research Project, 2018.

Howard, Philip N., Bharath Ganesh, Dimitra Liotsiou, John Kelly, and Camille François. "The IRA, Social Media and Political Polarization in the United States, 2012–2018 (Appendices)." Oxford: Computational Propaganda Research Project, 2018.

Hughes, Thomas P. *Rescuing Prometheus: Four Monumental Projects That Changed the Modern World*. New York: Pantheon Books, 1998.

Humes, Edward. *Garbology: Our Dirty Love Affair with Trash*. New York: Penguin, 2013.

Illouz, Eva. *Saving the Modern Soul: Therapy, Emotions, and the Culture of Self-Help*. Berkeley: University of California Press, 2008.

Insikt Group. "The Price of Influence: Disinformation in the Private Sector." Recorded Future, September 30, 2019. https://www.recordedfuture.com/disinformation-service-campaigns/.

Jackson, Tim. "Live Better by Consuming Less?: Is There a 'Double Dividend' in Sustainable Consumption?" *Journal of Industrial Ecology* 9, no. 1–2 (2008): 19–36. https://doi.org/10.1162/1088198054084734.

Jamieson, Kathleen Hall. *Cyberwar: How Russian Hackers and Trolls Helped Elect a President: What We Don't, Can't, and Do Know*. New York: Oxford University Press, 2018.

The Jammer and Jack the Ripper. "The Official Phreaker's Manual," February 14, 1987. http://textfiles.com/phreak/PHREAKING/manual1.txt.

Jankowicz, Nina. "How an Anti-Trump Flash Mob Found Itself in the Middle of Russian Meddling." POLITICO, July 5, 2020. https://www.politico.com/news/magazine/2020/07/05/how-an-anti-trump-flash-mob-found-itself-in-the-middle-of-russian-meddling-348729.

Jansen, Sue Curry. "Semantic Tyranny: How Edward L. Bernays Stole Walter Lippmann's Mojo and Got Away With It and Why It Still Matters." *International Journal of Communication* 7 (April 30, 2013): 1094–1111.

Jeffries, Adrianne. "Dating Coach Shows How to Get Classified Military Intel Using Social Engineering." The Verge, August 4, 2013. https://www.theverge.com/2013/8/4/4585994/hacking-people-is-easy-a-dating-coach-shows-how-easy-it-is-to-get-classified-intel.

Jeffries, Ross. *Secrets of Speed Seduction—How To Create An Instantaneous Sexual Attraction In Any Woman You Meet.* Culver City, CA: Ross Jeffries, 1994. "Joint Statement from the ODNI, DOJ, FBI and DHS: Combating Foreign Influence in U.S. Elections." Washington, DC: Director of National Intelligence, October 19, 2018. https://www.dni.gov/index.php/newsroom/press-releases/item/1915-joint-statement-from-the-odni-doj-fbi-and-dhs-combating-foreign-influence-in-u-s-elections.

Jordan, John M. *Machine-Age Ideology: Social Engineering and American Liberalism, 1911–1939.* New ed. Chapel Hill, NC: The University of North Carolina Press, 2010.

Jordan, Tim, and Paul Taylor. "A Sociology of Hackers." *The Sociological Review*, January 25, 2017. https://journals.sagepub.com/doi/pdf/10.1111/1467-954X.00139.

Kahneman, Daniel. *Thinking, Fast and Slow.* New York: Farrar, Straus & Giroux, 2011.

Kaiser, Brittany. *Targeted: The Cambridge Analytica Whistleblower's Inside Story of How Big Data, Trump, and Facebook Broke Democracy and How It Can Happen Again.* New York: HarperCollins, 2019.

Karpf, Dave. "Will the Real Psychometric Targeters Please Stand Up?" Civic Hall, February 1, 2017. https://civichall.org/civicist/will-the-real-psychometric-targeters-please-stand-up/.

Keim, Megan Dunham. "Your Facts Are Not Safe With Us: Russian Information Operations As Social Engineering." Hacking Illustrated Series InfoSec Tutorial Videos. Presented at BSides, Philadelphia, 2017. Video, 50:28. https://www.irongeek.com/i.php?page=videos/bsidesphilly2017/bsidesphilly-cs01-your-facts-are-not-safe-with-us-russian-information-operations-as-social-engineering-meagan-dunham-keim.

Kellogg, Paul Underwood. *The Pittsburgh Survey: Findings in Six Volumes.* New York: Charities Publication Committee, 1910.

Kessler, Glenn. "Analysis | Foundation Faceoff: The Trump Foundation vs. the Clinton Foundation." *Washington Post*, June 27, 2018. https://www.washingtonpost.com/news/fact-checker/wp/2018/06/27/foundation-face-off-the-trump-foundation-versus-the-clinton-foundation/.

"Kevin Mitnick: Cyber Thief; Computer Hacker Is Caught and Sent to Prison for Invading Company Computers in What He Says Was Just a Challenge to Him." *60 Minutes*. CBS, January 23, 2000.

The Kid & Co. "How to Get into a C.O." *2600: The Hacker Quarterly* 2, no. 3 (March 1985).

The Kid & Co. and The Shadow. "More on Trashing: What to Look For, How to Act, Where to Go." *2600: The Hacker Quarterly* 1, no. 9 (September 1984). http://textfiles.com/phreak/TRASHING/trash.phk.

King, Andrew Stephen. "Feminism's Flip Side: A Cultural History of the Pickup Artist." *Sexuality & Culture* 22, no. 1 (March 1, 2018): 299–315. https://doi.org/10.1007/s12119-017-9468-0.

Kirschenbaum, Matthew G. *Mechanisms: New Media and the Forensic Imagination*. Cambridge, MA: MIT Press, 2012.

Kluepfel, Henry M. "Foiling the Wiley Hacker: More than Analysis and Containment." In *Proceedings. International Carnahan Conference on Security Technology*, 15–21, 1989. https://doi.org/10.1109/CCST.1989.751947.

Konkin III, Samuel Edward. "New Libertarian Manifesto." Huntington Beach, CA: Koman Publishing, 1983. http://agorism.info/docs/NewLibertarianManifesto.pdf.

Koopman, Colin. *How We Became Our Data: A Genealogy of the Informational Person*. Chicago: University of Chicago Press, 2019.

Krenn, Mario. "From Scientific Management to Homemaking: Lillian M. Gilbreth's Contributions to the Development of Management Thought." *Management & Organizational History* 6, no. 2 (May 1, 2011). https://doi.org/10.1177/1744935910397035.

Lamme, Margot Opdycke. "Outside the Prickly Nest: Revisiting Doris Fleischman." *American Journalism* 24, no. 3 (July 1, 2007): 85–107. https://doi.org/10.1080/08821127.2007.10678080.

Langer, Ellen J. *Mindfulness*. 25th anniversary ed. Boston: Da Capo Lifelong Books, 2014.

Langer, Ellen J. *The Power of Mindful Learning*. Reprint ed. Boston: Da Capo Lifelong Books, 2016.

Lanier, Tripp. "Jordan Harbinger of Pickup Podcast: How to Be Charismatic (episode 094)." April 26, 2010. In The New Man Podcast. MP3 audio, 30:54. http://www.thenewmanpodcast.com/2010/04/jordan-harbinger-charisma/.

Lapowsky, Issie. "25 Geniuses Who Are Creating the Future of Business." *Wired*, April 26, 2016. https://www.wired.com/2016/04/wired-nextlist-2016/.

Lapsley, Phil. *Exploding the Phone: The Untold Story of the Teenagers and Outlaws Who Hacked Ma Bell*. New York: Grove, 2014.

Lasswell, Harold D. "The Theory of Political Propaganda." *American Political Science Review* 21, no. 3 (August 1927): 627–631. https://doi.org/10.2307/1945515.

Lawson, Sean. "Beyond Cyber-Doom: Assessing the Limits of Hypothetical Scenarios in the Framing of Cyber-Threats." *Journal of Information Technology & Politics* 10, no. 1 (January 1, 2013): 86–103. https://doi.org/10.1080/19331681.2012.759059.

Lawson, Sean. "HBGary Hearts Apple." *Forbes.com*, February 22, 2011.

Lawson, Sean. "Putting the 'War' in Cyberwar: Metaphor, Analogy, and Cybersecurity Discourse in the United States." *First Monday* 17, no. 7 (2012). https://doi.org/10.5210/fm.v17i7.3848.

Lawson, Sean T. *Cybersecurity Discourse in the United States: Cyber-Doom Rhetoric and Beyond*. London: Routledge, 2020.

Le Bon, Gustave. *The Crowd: A Study of the Popular Mind*. London: Ernest Benn, 1896.

Lee, Gerald Stanley. *Crowds: A Moving-Picture of Democracy*. Garden City, NY: Doubleday Page, 1913. http://www.gutenberg.org/ebooks/15759.

Lee, Ralph, dir. *The Secret History of Hacking*. Produced by Mira King. September Films, 2001. Video, 50:08. Uploaded to YouTube in 2013, https://www.youtube.com/watch?v=PUf1d-GuK0Q.

Lee, Ivy Ledbetter. *Human Nature and the Railroads*. E. S. Nash, 1915. http://archive.org/details/humannatureandr01leegoog.

Lepore, Jill. "Not So Fast." *New Yorker*, October 5, 2009. https://www.newyorker.com/magazine/2009/10/12/not-so-fast.

Levine, Timothy R., Kim B. Serota, Hillary Shulman, David D. Clare, Hee Sun Park, Allison S. Shaw, Jae Chul Shim, and Jung Hyon Lee. "Sender Demeanor: Individual Differences in Sender Believability Have a Powerful Impact on Deception Detection Judgments." *Human Communication Research* 37, no. 3 (2011): 377–403.

Levy, Steven. "The Case for Hackers." *Newsweek*, February 6, 1995.

Lewis, Justin, and Sut Jhally. "The Struggle Over Media Literacy." *Journal of Communication* 48, no. 1 (1998): 109–120. https://doi.org/10.1111/j.1460-2466.1998.tb02741.x.

Linde, Richard R. "Operating System Penetration." In *AFIPS '75: Proceedings of the May 19–22, 1975, National Computer Conference and Exposition*, 361–368. New York: Association for Computing Machinery, 1975. https://doi.org/10.1145/1499949.1500018.

Lingel, Jessa. "Adjusting the Borders: Bisexual Passing and Queer Theory." *Journal of Bisexuality* 9, no. 3–4 (2009): 381–405. https://doi.org/10.1080/15299710903316646.

Linvill, Darren L., Brandon C. Boatwright, Will J. Grant, and Patrick L. Warren. "'The Russians Are Hacking My Brain!' Investigating Russia's Internet Research Agency Twitter Tactics During the 2016 United States Presidential Campaign." *Computers in Human Behavior* 99 (October 1, 2019): 292–300. https://doi.org/10.1016/j.chb.2019.05.027.

Lippmann, Walter. *Drift and Mastery: An Attempt to Diagnose the Current Unrest*. New York: Mitchell Kennerley, 1914.

Lippmann, Walter. *Public Opinion*. New York: Harcourt, Brace, 1922.

Littman, Jonathan. "Most Wanted: In the Mind of Hacker Kevin Mitnick." *Computerworld*, January 15, 1996. https://www.computerworld.com/article/2531739/most-wanted.html.

Littman, Jonathan. *The Fugitive Game: Online with Kevin Mitnick*. Boston: Little, Brown, 1997.

Long, Johnny. *No Tech Hacking: A Guide to Social Engineering, Dumpster Diving, and Shoulder Surfing*. Edited by Scott Pinzon. Rockland, MA: Syngress, 2008.

Lumley, Frederick Elmore. *The Propaganda Menace, By Frederick E. Lumley.* New York: Century, 1933.

Lynch, Stephen. "Kevin Mitnick Reboots His Life // For the World's Most Notorious Computer Hacker, the Hardest Thing to Crack Is His Own Reputation." *Orange County Register*, November 8, 2001.

Maccoby, Nathan, Freddie O. Sabghir, and Bryant Cushing. "A Method for the Analysis of the News Coverage of Industry." *Public Opinion Quarterly* 14, no. 4 (January 1, 1950): 753–758. https://doi.org/10.1086/266253.

Makuch, Ben and "Jek." "The Penetration Tester." October 15, 2020. In CYBER (podcast). MP3 audio, 29:00. https://podcasts.apple.com/us/podcast/re-run-the-penetration-tester/id1441708044?i=1000494881794.

Margolin, Leo Jay. *Paper Bullets: A Brief Story of Psychological Warfare in World War II.* New York: Froben Press, 1946.

Marine, Gene. *America the Raped: The Engineering Mentality and the Devastation of a Continent.* New York: Simon & Schuster, 1969.

Markoff, John. "Taking a Computer Crime to Heart." *New York Times*, January 28, 1995. https://www.nytimes.com/1995/01/28/business/taking-a-computer-crime-to-heart.html.

Marshall, Jennifer Jane. "Clean Cuts: Procter & Gamble's Depression-Era Soap-Carving Contests." *Winterthur Portfolio* 42, no. 1 (2008): 51–76. https://doi.org/10.1086/528905.

"Master Hacker Kevin Mitnick Shares His 'Addiction.'" August 21, 2011. NPR. MP3 audio, 7:49. https://www.npr.org/2011/08/21/139677992/master-hacker-kevin-mitnick-shares-his-addiction.

Mattelart, Armand. *Networking the World, 1794–2000.* English language ed. Translated by Liz Carey-Libbrecht and James A. Cohen. Minneapolis: University of Minnesota Press, 2000.

Matteson, Miriam L., Lorien Anderson, and Cynthia Boyden. "'Soft Skills': A Phrase in Search of Meaning." *Portal: Libraries and the Academy* 16, no. 1 (February 18, 2016): 71–88. https://doi.org/10.1353/pla.2016.0009.

Maxwell, Richard, and Toby Miller. *Greening the Media.* New York: Oxford University Press, 2012.

McClain, Noah. "Caught Inside the Black Box: Criminalization, Opaque Technology, and the New York Subway MetroCard." *The Information*

Society: An International Journal 35, no. 5 (July 25, 2019): 251–271. https://doi.org/10.1080/01972243.2019.1644410.

McClymer, John F. *War and Welfare: Social Engineering in America, 1890–1925*. Westport, CT.: Greenwood Press, 1980.

McRobbie, Angela. *The Uses of Cultural Studies: A Textbook*. London: SAGE, 2005.

Mears, Daniel P. "The Ubiquity, Functions, and Contexts of Bullshitting." *Journal of Mundane Behavior* 3, no. 2 (June 2002): 233–256.

Melendez, Steven. "Florida Democrats Get Threatening Emails Demanding That They Vote for Trump." *Fast Company*, October 20, 2020. https://www.fastcompany.com/90566468/florida-democrats-get-threatening-emails-demanding-that-they-vote-for-trump.

Melosi, Martin V. *Garbage in the Cities: Refuse, Reform, and the Environment*. Revised ed. History of the Urban Environment. Pittsburgh: University of Pittsburgh Press, 2005.

The Mentor. "Engineering." Phoenix Project BBS, July 13, 1988. http://www.textfiles.com/messages/phoenix1.msg.

Miao, Hannah. "Democratic Senators Formally Request Investigation of Hawley and Cruz after Deadly Capitol Insurrection." CNBC, January 21, 2021. https://www.cnbc.com/2021/01/21/capitol-riot-democrats-file-ethics-complaint-against-cruz-hawley.html.

Michaels, David. *The Triumph of Doubt: Dark Money and the Science of Deception*. New York: Oxford University Press, 2020.

Michaels, Spenser. "The Philadelphia Story, Part 1." *TAP*, August 1982. http://www.textfiles.com/tap/issue.76/tap7602.jpg.

Mills, C. Wright. *The Sociological Imagination*. New York: Oxford University Press, 1959.

Mindell, David A. "'The Clangor of That Blacksmith's Fray': Technology, War, and Experience Aboard the USS Monitor." *Technology and Culture* 36, no. 2 (1995): 242–270. https://doi.org/10.2307/3106372.

Mistry, Kaeten. "The Case for Political Warfare: Strategy, Organization and US Involvement in the 1948 Italian Election." *Cold War History* 6, no. 3 (2006): 301–329. https://doi.org/10.1080/14682740600795451.

Mitnick, Kevin. "Kevin Mitnick's Testimony to Senate Homeland Security and Government Affairs Committee." *Cyber Attack: Is the Government Safe?* Governmental Affairs Committee (March 2, 2000). https://www.hsgac.senate.gov/imo/media/doc/mitnick.pdf.

Mitnick, Kevin D. "Kevin Mitnick—'I'm The Hacker Who Changed Sides'; First Person." *Financial Times*, May 30, 2009.

Mitnick, Kevin D. *The Art of Deception: Controlling the Human Element of Security*. Hoboken: Wiley, 2002.

Mitnick, Kevin D., and William L. Simon. *Ghost in the Wires: My Adventures as the World's Most Wanted Hacker*. London: Little, Brown, 2012.

Mohun, Arwen. "Industrial Genders: Home/Factory." In *Gender & Technology: A Reader*, edited by Nina E. Lerman, Ruth Oldenziel, and Arwen P. Mohun, 153–176. Baltimore: Johns Hopkins University Press, 2003.

Monberg, John, and Stuart L. Esrock. "What a Long, Strange Trip It's Been: The Past, Present, and Uncertain Future of Universal Service." *Convergence: The International Journal of Research into New Media Technologies* 6, no. 4 (December 1, 2000): 78–92. https://doi.org/10.1177/135485650000600406.

Morley, David, and Kevin Robins. "Techno-Orientalism: Japan Panic." Chap. 8 in *Spaces of Identity: Global Media, Electronic Landscapes and Cultural Boundaries*. New York: Routledge, 1995.

Mostegel, Iris. "The Original Influencer." *History Today*, February 6, 2019. https://www.historytoday.com/miscellanies/original-influencer.

Mueller, Robert. "Report On The Investigation Into Russian Interference In The 2016 Presidential Election." Washington, DC: Department of Justice, March 2019. https://cdn.cnn.com/cnn/2019/images/04/18/mueller-report-searchable.pdf.

Mukerji, Chandra. "Bullshitting: Road Lore Among Hitchhikers." *Social Problems* 25, no. 3 (February 1, 1978): 241–252. https://doi.org/10.2307/800062.

Murphree, Vanessa. "Edward Bernays's 1929 'Torches of Freedom' March: Myths and Historical Significance." *American Journalism* 32, no. 3 (July 3, 2015): 258–281. https://doi.org/10.1080/08821127.2015.1064681.

Nakashima, Ellen. "U.S. Cyber Command Operation Disrupted Internet Access of Russian Troll Factory on Day of 2018 Midterms." *Washington*

Post, February 27, 2019. https://www.washingtonpost.com/world/national-security/us-cyber-command-operation-disrupted-internet-access-of-russian-troll-factory-on-day-of-2018-midterms/2019/02/26/1827fc9e-36d6-11e9-af5b-b51b7ff322e9_story.html.

Nakashima, Ellen, Amy Gardner, and Aaron C. Davis. "FBI Links Iran to Online Hit List Targeting Top Officials Who've Refuted Trump's Election Fraud Claims." *Washington Post*, December 22, 2020. https://www.washingtonpost.com/national-security/iran-election-fraud-violence/2020/12/22/4a28e9ba-44a8-11eb-a277-49a6d1f9dff1_story.html.

Nakashima, Ellen, Amy Gardner, Isaac Stanley-Becker, and Craig Timberg. "U.S. Government Concludes Iran Was behind Threatening Emails Sent to Democrats." *Washington Post*, October 22, 2020. https://www.washingtonpost.com/technology/2020/10/20/proud-boys-emails-florida/.

Nakashima, Ellen, and David L. Stern. "U.S. Sanctions Ukrainians Involved in Russia-Linked Campaign Promoted by Giuliani to Smear Biden." *Washington Post*, January 11, 2021. https://www.washingtonpost.com/national-security/ukranians-sanctions-giuliani-election-interference/2021/01/11/0c447aea-5436-11eb-a08b-f1381ef3d207_story.html.

Navarro, Joe, and Marvin Karlins. *What Every BODY Is Saying: An Ex-FBI Agent's Guide to Speed-Reading People*. New York: William Morrow, 2008.

Newton, Casey. "Democrats Ran Influence Campaigns in at Least Three States During the Midterm Elections." *The Verge*, January 8, 2019. https://www.theverge.com/2019/1/8/18173027/democrats-misinformation-reid-hoffman-alabama-new-knowledge-influence-campaigns.

Nix, Alexander. "From Mad Men to Math Men." Presented at OMR Festival 2017, Hamburg, March 2017. Video, 30:17. https://www.youtube.com/watch?v=6bG5ps5KdDo.

Occupational Outlook Handbook. "Information Security Analysts." Bureau of Labor Statistics, April 9, 2021. https://www.bls.gov/ooh/computer-and-information-technology/information-security-analysts.htm.

Offenhuber, Dietmar. *Waste Is Information: Infrastructure Legibility and Governance*. Cambridge, MA: MIT Press, 2017.

Ono, Kent A. "'America's Apple Pie: Baseball, Japan-Bashing, and the Sexual Threat of Economic Miscegenation." In *Out of Bounds: Sports, Media, and the Politics of Identity*, edited by Aaron Baker and Todd Boyd, 81–101. Bloomington: University of Indiana Press, 1997.

Original Films of Frank B. Gilbreth, 1945. Video, 32:13. http://archive.org/details/0809_Original_Films_of_Frank_B_Gilbreth_02_12_34_00.

Orth, Maureen. "For Whom Ma Bell Tolls Not." *Los Angeles Times*, October 31, 1971.

Osterhout, Jacob E. "Stealth Marketing: When You're Being Pitched and You Don't Even Know It!" nydailynews.com, April 18, 2010. https://www.nydailynews.com/life-style/stealth-marketing-pitched-don-article-1.165278.

O'Sullivan, Donie, and Alex Marquardt. "Iranian Hackers Who Posed as the Proud Boys Accessed Voter Data in One State, Feds Say." CNN.com, October 31, 2020. https://www.cnn.com/2020/10/30/politics/iran-hackers-proud-boys/index.html.

O'Sullivan, Patrick B. "Bridging the Mass-Interpersonal Divide Synthesis Scholarship in HCR." *Human Communication Research* 25, no. 4 (1999): 569–588. https://doi.org/10.1111/j.1468–2958.1999.tb00462.x.

O'Sullivan, Patrick B., and Caleb T. Carr. "Masspersonal Communication: A Model Bridging the Mass-Interpersonal Divide." *New Media & Society* 20, no. 3 (March 1, 2018): 1161–1180. https://doi.org/10.1177/1461444816686104.

Packard, Vance. *The Hidden Persuaders*. London: Longmans, Green & Company, 1957.

Packard, Vance. *The Waste Makers*. Philadelphia: David McKay, 1960.

Palenchar, Michael J., and Kathy R. Fitzpatrick. "Secret Persuaders: Ethical and Rhetorical Perspectives on the Use of Public Relations Front Groups." In *Rhetorical and Critical Approaches to Public Relations II*, edited by Robert L. Heath, Elizabeth L. Toth, and Damion Waymer, 272–289. New York: Routledge, 2009.

Parks, Miles. "Florida Governor Says Russian Hackers Breached 2 Counties In 2016." NPR.org, May 14, 2019. https://www.npr.org/2019/05/14/723215498/florida-governor-says-russian-hackers-breached-two-florida-counties-in-2016.

Paul, Christopher, and Miriam Matthews. "The Russian 'Firehose of Falsehood' Propaganda Model: Why It Might Work and Options to Counter It." Santa Monica, CA: RAND Corporation, 2016. https://doi.org/10.7249/PE198.

"Penetration Tester Job Description." JobHero. Accessed August 16, 2020. https://www.jobhero.com/job-description/examples/information-technology/penetration-tester.

Pfeffer, Carla A. "'I Don't like Passing as a Straight Woman': Queer Negotiations of Identity and Social Group Membership." *American Journal of Sociology* 120, no. 1 (2014): 1–44. https://doi.org/10.1086/677197.

Phillips, Kevin. *Mediacracy: American Parties and Politics in the Communications Age.* Garden City, NY: Doubleday, 1975.

Phillips, Whitney. *This Is Why We Can't Have Nice Things: Mapping the Relationship between Online Trolling and Mainstream Culture.* Cambridge, MA: MIT Press, 2015.

"Phunware Announces Strategic Relationship with American Made Media Consultants for the Trump–Pence 2020 Reelection Campaign's Mobile Application Portfolio." Phunware, May 27, 2020. https://web.archive.org/web/20200926083742/https://www.phunware.com/press-releases/american-made-consultants-trump-pence-2020-relelection-campaign-mobile-application-portfolio/.

"Phunware Data Licensing." Phunware, accessed September 19, 2020. https://web.archive.org/web/20200919010301/https://www.phunware.com/data/data-licensing/.

Pomerantsev, Peter. *Nothing Is True and Everything Is Possible: The Surreal Heart of the New Russia.* New York: PublicAffairs, 2014.

Pomerantsev, Peter. *This Is Not Propaganda: Adventures in the War Against Reality.* New York: PublicAffairs, 2019.

Pompeo, Michael R. "Sanctioning Russia-Linked Disinformation Network for Its Involvement in Attempts to Influence U.S. Election." *US Department of State* (blog), January 11, 2021. https://2017–2021.state.gov/sanctioning-russia-linked-disinformation-network-for-its-involvement-in-attempts-to-influence-u-s-election/index.html.

Popken, Ben. "Trolls for Hire: Russia's Freelance Disinformation Firms Offer Propaganda with a Professional Touch." NBC News, October 1, 2019. https://www.nbcnews.com/tech/security/trolls-hire-russia-s-freelance-disinformation-firms-offer-propaganda-professional-n1060781.

Porat, Aelon. "Power to the People." Presented at HOPE, June 26, 2020. https://infocondb.org/con/hope/hope-2020/power-to-the-people-effective-advocacy-for-privacy-and-security.

"Program Details » MS in Cybersecurity Program Overview." The University of Arizona, accessed March 13, 2020. https://cybersecurity.arizona.edu/program/.

Pruitt, Gary. "Putin Says Russia Didn't Meddle in US Vote, despite Evidence." *AP News*, June 6, 2019. https://apnews.com/article/1f12eaf734014da6bcd0995ae89c6636.

Quann, John, and Peter Belford. "The Hack Attack—Increasing Computer System Awareness of Vulnerability Threats." In *3rd Applying Technology to Systems; Aerospace Computer Security Conference*, 155–157. Orlando: American Institute of Aeronautics and Astronautics, 1987. https://doi.org/10.2514/6.1987-3093.

Rake, Derek. "NLP Seduction Patterns, Routines, Phrases & Scripts." Derek Rake (blog), June 29, 2018. https://derekrake.com/blog/nlp-seduction-patterns/.

Ramasubramanian, Srividya, and Omotayo O. Banjo. "Critical Media Effects Framework: Bridging Critical Cultural Communication and Media Effects through Power, Intersectionality, Context, and Agency." *Journal of Communication* 70, no. 3 (June 1, 2020): 379–400. https://doi.org/10.1093/joc/jqaa014.

Reagle, Jr., Joseph M. *Hacking Life: Systematized Living and Its Discontents*. Cambridge, MA: MIT Press, 2019.

Reeves, Jay, Lisa Mascaro, Calvin Woodward, Dustin Weaver, and Michael Casey. "Capitol Riot More Sinister than It Looked as Gallows, Pipes and Guns Turn Up." *Tampa Bay Times*, January 11, 2021. https://www.tampabay.com/news/florida-politics/2021/01/11/capitol-riot-more-sinister-that-it-looked-as-gallows-pipes-and-guns-turn-up/.

"Revealed: Trump Campaign Strategy to Deter Millions of Black Americans from Voting in 2016." *Channel 4 News*, 2020. Video, 21:39. https://www.youtube.com/watch?v=KIf5ELaOjOk.

Reynolds, John. "Trashing the Phone Company." *Telephone Electronics Line* 2, no. 5 (May 1975). http://pdf.textfiles.com/zines/TEL/tel-07.pdf.

Rid, Thomas. *Active Measures: The Secret History of Disinformation and Political Warfare*. New York: Farrar, Straus and Giroux, 2020.

Rid, Thomas. "Insisting That the Hunter Biden Laptop Is Fake Is a Trap. So Is Insisting That It's Real." *Washington Post*, October 24, 2020. https://

www.washingtonpost.com/outlook/2020/10/24/hunter-biden-laptop-disinformation/.

Root, Alfred R., and Alfred C. Welch. "The Continuing Consumer Study: A Basic Method for the Engineering of Advertising." *Journal of Marketing* 7, no. 1 (July 1942): 3–21. https://doi.org/10.2307/1246447.

Rosenberg, Matthew, Nicholas Confessore, and Carole Cadwalladr. "How Trump Consultants Exploited the Facebook Data of Millions." *New York Times*, March 17, 2018. https://www.nytimes.com/2018/03/17/us/politics/cambridge-analytica-trump-campaign.html.

Rosenberg, Matthew, Nicole Perlroth, and David E. Sanger. "'Chaos Is the Point': Russian Hackers and Trolls Grow Stealthier in 2020." *New York Times*, January 10, 2020. https://www.nytimes.com/2020/01/10/us/politics/russia-hacking-disinformation-election.html.

Rosenfeld, Megan. "At Surveillance Expo, Sneak Peeks at the Sweet Spy and Buy." *Washington Post*. December 18, 1989. https://www.washingtonpost.com/archive/business/1989/12/18/at-surveillance-expo-sneak-peeks-at-the-sweet-spy-and-buy/4b130402-43b1-4942-ada7-df34de8e51de/.

Rosenstein, Rod J. "Report of the Attorney General's Cyber Digital Task Force." US Department of Justice, 2018. https://www.justice.gov/archives/ag/page/file/1076696/download.

Ruck, Damian J., Natalie M. Rice, Joshua Borycz, and R. Alexander Bentley. "Internet Research Agency Twitter Activity Predicted 2016 U.S. Election Polls." *First Monday* 24, no. 7 (June 30, 2019). https://doi.org/10.5210/fm.v24i7.10107.

Russell, Jim. "Sorry, the Telephone Company You're Dialing Has Been Temporarily Disconnected." Washington, DC: National Public Radio, January 1973. http://explodingthephone.com/docs/dbx0659.pdf.

Russell, Karen Miller, and Carl O. Bishop. "Understanding Ivy Lee's Declaration of Principles: U.S. Newspaper and Magazine Coverage of Publicity and Press Agentry, 1865–1904." *Public Relations Review* 35, no. 2 (June 2009): 91–101. https://doi.org/10.1016/j.pubrev.2009.01.004.

"Russian Active Measures Campaigns and Interference in the 2016 U.S. Election. Volume 2: Russia's Use of Social Media With Additional Views." Washington, DC: US Senate Committee on Intelligence, 2019. https://

www.intelligence.senate.gov/sites/default/files/documents/Report_Volume2.pdf.

"Russian National Charged with Interfering in U.S. Political System." US Department of Justice, October 19, 2018. https://www.justice.gov/opa/pr/russian-national-charged-interfering-us-political-system.

Ryan, Thomas, and Gabriella Mauch. "Getting in Bed with Robin Sage." Provide Security and BlackHat USA, 2010.

Ryan-Mosley, Tate. "The Technology That Powers the 2020 Campaigns, Explained." *MIT Technology Review*, September 28, 2020. https://www.technologyreview.com/2020/09/28/1008994/the-technology-that-powers-political-campaigns-in-2020-explained/.

Satin, Mark. *New Age Politics: The Emerging New Alternative to Liberalism and Marxism*. Gearhart, OR: Fairweather Press, 1976.

Satter, Raphael, Jeff Donn, and Chad Day. "Inside Story: How Russians Hacked the Democrats' Emails." AP News, November 4, 2017. https://apnews.com/dea73efc01594839957c3c9a6c962b8a.

Sauter, Molly R. "Kevin Mitnick, *New York Times*, and the Media's Conception of the Hacker." In *Making Our World: The Hacker and Maker Movements in Context*, edited by Jeremy Hunsinger and Andrew Schrock, 21–35. New York: Peter Lang, 2019.

Scanlan, John. *On Garbage*. London: Reaktion, 2005.

"Scenic Drive." Takedown, 1995. http://www.takedown.com/evidence/transcripts/best.html.

Schechner, Sam, Emily Glazer, and Patience Haggin. "Political Campaigns Know Where You've Been. They're Tracking Your Phone." *Wall Street Journal*, October 10, 2019. https://www.wsj.com/articles/political-campaigns-track-cellphones-to-identify-and-target-individual-voters-11570718889.

Schmitt, Michael N., ed. *Tallinn Manual 2.0 on the International Law Applicable to Cyber Operations*. 2nd ed. Cambridge: Cambridge University Press, 2017.

Schneier, Bruce. "Toward an Information Operations Kill Chain." Lawfare, April 24, 2019. https://www.lawfareblog.com/toward-information-operations-kill-chain.

Schneier, Bruce. "The US Has Suffered a Massive Cyberbreach. It's Hard to Overstate How Bad It Is." *Guardian*, December 23, 2020. http://www.theguardian.com/commentisfree/2020/dec/23/cyber-attack-us-security-protocols.

Schultz, Stanley K., and Clay McShane. "To Engineer the Metropolis: Sewers, Sanitation, and City Planning in Late-Nineteenth-Century America." *The Journal of American History* 65, no. 2 (1978): 389–411. https://doi.org/10.2307/1894086.

Schweickert, Molly. "Cambridge Analytica explains how the Trump campaign worked." *D3con: The Future of Digital Advertising*, 2017. Video, 40:02. https://www.youtube.com/watch?v=bB2BJjMNXpA.Seligman, Lara. "Mattis Confirms Russia Interfered in U.S. Midterm Elections." *Foreign Policy*, December 1, 2018. https://foreignpolicy.com/2018/12/01/mattis-confirms-russia-interfered-in-us-midterm-elections-putin-trump/.

Serres, Michel. *The Parasite*. Translated by Lawrence R. Schehr. Baltimore: Johns Hopkins University Press, 1982.

"Seven Commandments of Fake News—*New York Times* Exposes Kremlin's Methods." EUvsDisinfo, November 21, 2018. https://euvsdisinfo.eu/seven-commandments-of-fake-news-new-york-times-exposes-kremlins-methods/.

Shane, Scott, and Alan Blinder. "Democrats Faked Online Push to Outlaw Alcohol in Alabama Race." *New York Times*, January 7, 2019. https://www.nytimes.com/2019/01/07/us/politics/alabama-senate-facebook-roy-moore.html.

Shane, Scott, and Alan Blinder. "Secret Experiment in Alabama Senate Race Imitated Russian Tactics." *New York Times*, December 20, 2018. https://www.nytimes.com/2018/12/19/us/alabama-senate-roy-jones-russia.html.

Sharp Remob. "Sharp Remob's Guide to Bullshitting the Phone Company Out of Important Information." Archive. Textfiles.com, 1989. http://www.textfiles.com/phreak/soceng.txt.

Sherman, Justin, and Anastasios Arampatzis. "Social Engineering as a Threat to Societies: The Cambridge Analytica Case." RealClearDefense, July 18, 2018. https://www.realcleardefense.com/articles/2018/07/18/social_engineering_as_a_threat_to_societies_the_cambridge_analytica_case_113620.html.

Shimer, David. *Rigged: America, Russia, and One Hundred Years of Covert Electoral Interference*. New York: Knopf, 2020.

Shimomura, Tsutomu, and John Markoff. *Takedown: The Pursuit and Capture of Kevin Mitnick, America's Most Wanted Computer Outlaw—by the Man Who Did It*. New York: Hyperion, 1996.

Silverman, Craig, Jane Lytvynenko, and William Kung. "Disinformation For Hire: How A New Breed Of PR Firms Is Selling Lies Online." *BuzzFeed News*, January 6, 2020. https://www.buzzfeednews.com/article/craigsilverman/disinformation-for-hire-black-pr-firms.

Simpson, Christopher. *Science of Coercion: Communication Research and Psychological Warfare, 1945–1960*. New York: Oxford University Press, 1994.

Slatalla, Michelle, and Joshua Quittner. *Masters of Deception: The Gang That Ruled Cyberspace*. New York: HarperCollins, 1995. http://archive.org/details/mastersofdecepti00slat.

Slayton, Rebecca. "The Paradoxical Authority of the Certified Ethical Hacker." Limn, February 14, 2017. https://limn.it/articles/the-paradoxical-authority-of-the-certified-ethical-hacker/.

Smith, Adam. *The Wealth of Nations*. New York: Bantam Classics, 2003.

Smith, Bruce Lannes. "Propaganda Analysis and the Science of Democracy." *Public Opinion Quarterly* 5, no. 2 (June 1941): 250–259. https://doi.org/10.1086/265490.

SN, Chesire Catalyst, and Emmanuel Goldstein. "Hackers On Planet Earth: Social Engineering." *2600: The Hacker Quarterly*, 1994. Video, 1:00:02. http://archive.org/details/HOPE-1-Social_Engineering.

Snyder, Timothy. *The Road to Unfreedom: Russia, Europe, America*. Reprint Edition. New York: Tim Duggan, 2019.

Social Engineering in Cincinnati: The Annual Report of the Council of Social Agencies of Cincinnati. Cincinnati, OH: The Council of Social Agencies of Cincinnati, 1919.

Social-Engineer—Professional Social Engineering Training and Services. "2-Day Social Engineering Bootcamp." Accessed March 13, 2020. https://web.archive.org/web/20170407134747/https://www.social-engineer.com/2-day-social-engineering-bootcamp/.

Social-Engineer. "The Human Hacking Conference 2021: Plan to Be Amazed!" Security Through Education (blog), June 22, 2020. https://www.social-engineer.org/social-engineering/the-human-hacking-conference-2021-plan-to-be-amazed/.

Solovey, Mark. "Project Camelot and the 1960s Epistemological Revolution: Rethinking the Politics-Patronage-Social Science Nexus." *Social Studies of Science* 31, no. 2 (2001): 171–206. https://doi.org/10.1177/0306312701031002003.

Solovey, Mark. "Science and the State During the Cold War: Blurred Boundaries and a Contested Legacy." *Social Studies of Science* 31, no. 2 (2001): 165–170. https://doi.org/10.1177/0306312701031002002.

"Some Thoughts on 'Garbage Picking.'" *2600: The Hacker Quarterly*, February 1984.

Sotomayor, Marianna. "Biden Campaign's Microtargeting of Latino Communities Takes on a New Twist." NBC News, October 13, 2020. https://www.nbcnews.com/politics/2020-election/biden-campaign-s-microtargeting-latino-communities-takes-new-twist-n1243170.

Spelman, Elizabeth V. "Combing Through the Trash: Philosophy Goes Rummaging." *The Massachusetts Review* 52, no. 2 (2011): 313–325.

Spenser, Edmund. "The Faerie Queene." The Faerie Queene, 1995. https://scholarsbank.uoregon.edu/xmlui/bitstream/handle/1794/784/faeriequeene.pdf.

Sproule, J. Michael. "Progressive Propaganda Critics and the Magic Bullet Myth." *Critical Studies in Mass Communication* 6, no. 3 (September 1989): 225–246. https://doi.org/10.1080/15295038909366750.

Stanford Internet Observatory. *Reply-Guys Go Hunting: An Investigation into a U.S. Astroturfing Operation on Facebook, Twitter, and Instagram*. Palo Alto, CA: Stanford University, 2020. https://cyber.fsi.stanford.edu/io/news/oct-2020-fb-rally-forge.

Stanley-Becker, Isaac. "Facebook Bans Marketing Firm Running 'Troll Farm' for Pro-Trump Youth Group." *Washington Post*, October 8, 2020. https://www.washingtonpost.com/technology/2020/10/08/facebook-bans-media-consultancy-running-troll-farm-pro-trump-youth-group/.

Starr, Paul. *The Creation of the Media: Political Origins of Modern Communication*. New York: Basic Books, 2005.

"Statement by NCSC Director William Evanina: Election Threat Update for the American Public." Office of the Director of National Intelligence, August 7, 2020. https://www.dni.gov/index.php/newsroom/press-releases/item/2139-statement-by-ncsc-director-william-evanina-election-threat-update-for-the-american-public.

Steinbeck, Fred. "Dealing with the Rate & Route Operator." *TAP*, November 1983. http://www.textfiles.com/tap/issue.88/tap8803.jpg.

Steinbeck, Fred. "Gibberish." *TAP*, November 1982. http://www.textfiles.com/tap/issue.79/tap7903.jpg.

Steinbeck, Fred. "Verification." *TAP*, November 1983. http://www.textfiles.com/tap/issue.88/tap8804.jpg.

Stokke, Andreas, and Don Fallis. "Bullshitting, Lying, and Indifference toward Truth." *Ergo, an Open Access Journal of Philosophy* 4, no. 10 (2017). http://dx.doi.org/10.3998/ergo.12405314.0004.010.

Stolper, Harold, and Jeff Jones. "The Crime of Being Short $2.75: Policing Communities of Color at the Turnstile." New York: Community Service Society, October 2017. https://www.cssny.org/publications/entry/the-crime-of-being-short-2.75.

Street, Jayson E. "DEFCON 19: Steal Everything, Kill Everyone, Cause Total Financial Ruin! (W Speaker)." Las Vegas, August 2011. https://www.youtube.com/watch?v=JsVtHqICeKE.

Takahashi, Dean. "Kevin Mitnick: An Interview on Trump, Russians, and Blockchain with the World's Most Famous Hacker (Updated)." VentureBeat (blog), July 28, 2018. https://venturebeat.com/2018/07/28/kevin-mitnick-an-interview-on-trump-russians-and-blockchain-with-the-worlds-most-famous-hacker/.

Taranto, James. "The Right's Happy Warrior." *Wall Street Journal*, April 30, 2010.

Tarbell, Ida M. *The History of the Standard Oil Company*. New York: McClure, Phillips & Co., 1904.

Temperton, James. "Ashley Madison Data Posted Online—and It's Worse than Anyone Thought." *Wired UK*, August 19, 2015. http://www.wired.co.uk/news/archive/2015-08/19/ashley-madison-hack-data-leaked-online.

"Testimony of Susan Headley, Tujunga, Calif." In *Computer Security in the Federal Government and the Private Sector: Hearings before the Subcommittee* on Oversight of Government Management of the Committee on Governmental Affairs, United States Senate, Ninety-Eighth Congress, First Session, October 25 and 26, 1983, 22–29. Washington, DC: Government Printing Office, 1983. http://hdl.handle.net/2027/pst.000012047208.

"The Electronic Delinquents." *20/20*. ABC, April 22, 1982.

"The SEORG Book List." Security Through Education, August 14, 2019. https://www.social-engineer.org/resources/seorg-book-list/.

Thibault, Ghislain. "Needles and Bullets: Media Theory, Medicine, and Propaganda, 1910–1940." In *Endemic: Essays in Contagion Theory*, edited by Kari Nixon and Lorenzo Servitje, 67–92. London: Palgrave Macmillan, 2016. https://doi.org/10.1057/978-1-137-52141-5_4.

Thomas, Douglas. *Hacker Culture*. Minneapolis: University of Minnesota Press, 2002.

Thornton, Mary. "Even Checking Lip Prints Wouldn't Offer Absolute Protection." *Washington Post*. May 22, 1984.

Thylstrup, Nanna Bonde. "Data Out of Place: Toxic Traces and the Politics of Recycling." *Big Data & Society* 6, no. 2 (July 1, 2019). https://doi.org/10.1177/2053951719875479.

Timberg, Craig, Tony Romm, Aaron C. Davis, and Elizabeth Dwoskin. "Secret Campaign to Use Russian-Inspired Tactics in 2017 Ala. Election Stirs Anxiety for Democrats." *Washington Post*, January 6, 2019. https://www.washingtonpost.com/business/technology/secret-campaign-to-use-russian-inspired-tactics-in-2017-alabama-election-stirs-anxiety-for-democrats/2019/01/06/58803f26-0400-11e9-8186-4ec26a485713_story.html.

Toffler, Alvin. *Future Shock*. New York: Bantam Books, 1970.

Toffler, Alvin, and Heidi Toffler. *Revolutionary Wealth*. New York:Knopf, 2006.

Tomes, Robert R. *U.S. Defense Strategy from Vietnam to Operation Iraqi Freedom: Military Innovation and the New American Way of War, 1973–2003*. London: Routledge, 2006.

Torres-Spelliscy, Ciara. "Dark Money as a Political Sovereignty Problem." *King's Law Journal* 28, no. 2 (May 4, 2017): 239–261. https://doi.org/10.1080/09615768.2017.1351659.

Tourish, Dennis. *Management Studies in Crisis: Fraud, Deception and Meaningless Research*. Cambridge: Cambridge University Press, 2019.

"Treasury Sanctions Russia-Linked Election Interference Actors." US Department of the Treasury, September 10, 2020. https://home.treasury.gov/news/press-releases/sm1118.

"Treasury Takes Further Action Against Russian-Linked Actors." US Department of the Treasury, January 11, 2021. https://home.treasury.gov/news/press-releases/sm1232.

Trottier, Daniel. "Open Source Intelligence, Social Media and Law Enforcement: Visions, Constraints and Critiques." *European Journal of Cultural Studies* 18, no. 4–5 (August 1, 2015): 530–547. https://doi.org/10.1177/1367549415577396.

Trueman, Charlotte. "What Impact Are Data Centres Having on Climate Change?" Computerworld, August 9, 2019. https://www.computerworld.com/article/3431148/why-data-centres-are-the-new-frontier-in-the-fight-against-climate-change.html.

Tufekci, Zeynep. "Engineering the Public: Big Data, Surveillance and Computational Politics." *First Monday* 19, no. 7 (July 2, 2014). http://www.firstmonday.dk/ojs/index.php/fm/article/view/4901.

Tullock, Gordon. *The Selected Works of Gordon Tullock*. Edited by Charles Rowley. Indianapolis: Liberty Fund, 2006.

Turner, Fred. *From Counterculture to Cyberculture: Stewart Brand, the Whole Earth Network, and the Rise of Digital Utopianism*. Chicago: University of Chicago Press, 2008.

Tye, Larry. *The Father of Spin: Edward L. Bernays and the Birth of Public Relations*. New York: Henry Holt, 2006.

Ueno, Toshiya. "Techno-Orientalism and Media-tribalism: On Japanese Animation and Rave Culture." *Third Text* 13, no. 47 (1999): 95–106. https://doi.org/10.1080/09528829908576801.

Umphress, David A. "Diving the Digital Dumpster: The Impact of the Internet on Collecting Open-Source Intelligence." *Air & Space Power Journal* 19, no. 4 (2005): 82–91.

United States Department of Justice. "United States vs. Internet Research Agency, et al.," February 16, 2018. https://www.justice.gov/file/1035477/download.

"US Report Finds No Direct Foreign Interference in 2018 Vote." *AP News*, December 21, 2018. https://apnews.com/article/cd2618aaeb6040c5b57fb 301361c76fd.

Valeriano, Brandon, and Ryan C. Maness. *Cyber War versus Cyber Realities: Cyber Conflict in the International System*. Oxford: Oxford University Press, 2015.

Verizon. *Data Breach Investigations Report*. 2018.

Verizon. *Data Breach Investigations Report*. 2019.

Vlamis, Kelsey. "A Texas Man Who Tweeted 'Assassinate AOC' Is Facing Charges in the Capitol Riot, FBI Says." Business Insider, January 24, 2021. https://www.businessinsider.com/texas-man-who-tweeted-assassinate-aoc -charged-in-capitol-riot-2021-1.

Volchek, Dmitry, and Daisy Sindelar. "One Professional Russian Troll Tells All." Radio Free Europe/Radio Liberty, March 25, 2015. https://www.rferl .org/a/how-to-guide-russian-trolling-trolls/26919999.html.

Wajcman, Judy. *Feminism Confronts Technology*. University Park, PA: Pennsylvania State University Press, 1991.

Wanless, Alicia, and Michael Berk. "Participatory Propaganda: The Engagement of Audiences in the Spread of Persuasive Communications." Presented at Social Media & Social Order, Culture Conflict 2.0, Oslo, November 2017. https://www.researchgate.net/profile/Alicia_Wanless/pub lication/329281610_Participatory_Propaganda_The_Engagement_of _Audiences_in_the_Spread_of_Persuasive_Communications/links/5c006 978299bf1a3c1561474/Participatory-Propaganda-The-Engagement-of -Audiences-in-the-Spread-of-Persuasive-Communications.pdf.

Watson, Gavin, Andrew Mason, and Richard Ackroyd. *Social Engineering Penetration Testing: Executing Social Engineering Pen Tests, Assessments and Defense*. Waltham, MA: Elsevier Science, 2014.

Watts, Clint. "Advanced Persistent Manipulators, Part Three: Social Media Kill Chain." Alliance for Securing Democracy (blog), July 22, 2019. https:// securingdemocracy.gmfus.org/advanced-persistent-manipulators-part -three-social-media-kill-chain/.

Watts, Clint. *Messing with the Enemy: Surviving in a Social Media World of Hackers, Terrorists, Russians, and Fake News*. New York: HarperCollins, 2018.

Webster, Stephen C. "Revealed: Air Force Ordered Software to Manage Army of Fake Virtual People." Raw Story, February 18, 2011.

Whittaker, Zack, and Natasha Lomas. "Even Years Later, Twitter Doesn't Delete Your Direct Messages." TechCrunch (blog), February 15, 2019. http://social.techcrunch.com/2019/02/15/twitter-direct-messages/.

Williams, Raymond. "Advertising: The Magic System." *Advertising & Society Review* 1, no. 1 (January 1, 2000). https://doi.org/10.1353/asr.2000.0016.

Wimberly, Cory. *How Propaganda Became Public Relations: Foucault and the Corporate Government of the Public*. New York: Routledge, 2020.

Winnet, Mike and Jenny Radcliffe. "Lies, Cons & People Hacking with Jenny Radcliffe (episode #5)." July 12, 2019. In *Not Another D*ckhead with a Podcast*. Video, 50:45. https://www.youtube.com/watch?v=qbKrK753wn0.

Wolfe, Jessica. "Nature and *Technê* in Spenser's *Faerie Queene*." In *A Companion to Tudor Literature*, edited by Kent Cartwright, 412–427. Hoboken: Blackwell, 2010.

Wolfe, Tom. "The 'Me' Decade and the Third Great Awakening." *New York Magazine*, August 23, 1976. https://nymag.com/news/features/45938/.

Won, Joseph D. "Yellowface Minstrelsy: Asian Martial Arts and the American Popular Imaginary." PhD diss., University of Michigan. ProQuest Dissertations Publishing, 1996. http://search.proquest.com/docview/304250501/?pq-origsite=primo.

Wortham, Jenna. "The Well-Followed on Social Media Cash In on Their Influence." *New York Times*, June 8, 2014. https://www.nytimes.com/2014/06/09/technology/stars-of-vine-and-instagram-get-advertising-deals.html.

Wylie, Christopher. *Mindf*ck: Cambridge Analytica and the Plot to Break America*. New York: Random House, 2019.

Yang, Guobin. "Communication as Translation: Notes toward a New Conception of Communication." In *Rethinking Media Research for Changing Societies*, edited by Matthew Powers and Adrienne Russell, 184–194. New York: Cambridge University Press, 2020.

Ybema, Sierk, Tom Keenoy, Cliff Oswick, Armin Beverungen, Nick Ellis, and Ida Sabelis. "Articulating Identities." *Human Relations* 62, no. 3 (March 1, 2009): 299–322. https://doi.org/10.1177/0018726708101904.

Index

Acxiom, 212
Adams, John, 151
Agorism, 53
Alt.seduction.fast, 143
American Council for Wider Reading, 94
American Statistical Association, 31–32
American Tobacco, 95–97, 135, 154
Andrejevic, Mark, 87, 212
Anonymous, 198
Apple Computers, 55
Art & Science of Dumpster Diving, The (Hoffman), 80
Art of Charm, The (podcast), 144
Art of Deception, The (Mitnick), 98, 147, 159
Ashley Madison, 85
Auerbach, Jonathan, 135, 152, 160
Authenticity, pretexting and, 92–93

Baker, Ray Stannard, 152
Bandler, Richard, 52
Bannon, Steve, 11, 171, 176, 181, 182–183, 184–185
Bazzell, Michael, 85, 212
Bell Telephone, 54–55, 128–129
Benjamin, Walter, 222

Benkler, Yochai, 202–203
Bernays, Edward
  American Tobacco and, 95–97, 135–136, 154
  on consent engineering, 157–158
  criticism of, 45
  Dixie Cups and, 75
  elitism and, 27–29, 62
  engineering of consent and, 11–12, 19, 39–42, 45–46
  as "father of public relations," 135–137
  field of public relations and, 37–38
  on influence on common man, 156
  "paper bullet" metaphor and, 152
  pretexts and, 22, 93–97
  stereotypes and, 89, 111–112, 113
  works of as references, 148
Biden, Hunter, 190–191
Biden, Joe, 190–191
Big Data, 72, 82–83
BIOC Agent 003, 130
Biography of an Idea (Bernays), 154
Blackberry, 210
Blackfist Facebook page, 186
Black Lives Matter, 3, 202
Black voting, suppression of, 205

"Blank," 99–101
Blue Boxes, 56, 117
Brand, Stewart, 51
Briant, Emma, 200, 210
Brown, Deborah, 126–128
Buchanan, James, 53
Buckley, William F., Jr., 43
Bulletin board systems (BBSs), 56
Bullet metaphor, 151–153, 187–188
Bullshitting
  among mass social engineers, 132–137
  Cambridge Analytica and, 180–185
  in hacker social engineering, 22, 130–132
  as indifference to truth, 118–120
  introduction to, 21, 115–116
  phone phreaks and, 58, 115, 117–118, 123–130
  potential responses to, 217–219
  Russia and, 177–180

Cadwalladr, Carole, 210
Cambridge Analytica
  assessments of, 11
  bullshitting and, 180–185
  continued impact of, 4–5
  effectiveness of, 202
  methods of, 12–13, 166–168
  microtargeting and, 3, 6–8, 14, 167
  penetrating and, 187–188
  pretexting and, 174–177
  trashing by, 171–172
Capitalism, transnational, 103–106
Capitol insurrection, 2
Carey, James, 153, 221, 222
Carnegie, Dale, 148
Carpenter, Perry, 103
Carr, Caleb, 16–17, 64, 167–168
Chesire Catalyst, 49
Cialdini, Robert, 148
Cigarette smoking, 95–97
Citizens United v. FEC, 215–216
Clark, A. B., 59–60

Clinton, Hillary, 2–3, 9–10, 202
Clinton Foundation, 181
Coercion, science of, 220, 221
Cognitive hacking, 6–7
Cold War, 157, 158–159, 198
Collins, Patricia Hill, 222–223
Color Fashion Bureau, 154
Committee for the Study and Protection of the Sanitary Dispensing of Foods and Drink, 94
Communication
  as domination, 153, 220, 224
  organizational structures and, 103–106
  strategic, 46, 198, 220
  as translation, 221–223
  transmission model of, 153, 221
Communication theory, 99
Computational Propaganda, 6
Computational Propaganda Project, 196–197
Conheady, Sharon
  on control, 145
  as ethical social engineer, 208, 211
  on hacker social engineering, 9
  on online data availability, 86
  on pretexting, 91, 103, 111, 112
  professionalization of field and, 63, 81, 148, 149–150
  on social engineer, 139
  on uses of social engineering, 159
Connell, Raewyn, 98–99
Consent engineers/engineering, 11–13, 19, 22, 39–40, 45–46, 63, 71, 75, 152, 157–158
Cooke, Morris, 34–35
Core dumps, 84
Craft spirit, loss of, 34–35
Credit documents, 75–76, 80
Credit ratings firms, 212
Creel, George, 152
Creel Committee, 152, 153
Crowd/masses, 31, 38–40, 151–152, 166

Crowds: A Moving-Picture of Democracy (Lee), 38–39
Cruz, Ted, 2, 175
Crystallizing Public Opinion (Bernays), 40–41
Cutlip, Scott, 43
Cyber kill chain, 221
Cyberlibertarian thinking, 51
Cyberspace Solarium Commission, 213–214, 220
Cyberwar (term), 7–8
Cyberwar: How Russian Hackers and Trolls Helped Elect a President (Jamieson), 7, 202

d3con, 188
Dark ads, 181
Dark money, 215–216
Data & Society, 217
Data dumps, 84–85
Data gathering/collection
  social engineering and, 33
  stricter laws on, 213–214
  Taylorists and, 35
  trashing and, 71–72
  unregulated, 212
Data retention, 215
"Declaration of Principles" (Lee), 115, 134, 135
DEF CON, 139, 150
Democratic National Committee, 2
De Payne, Lewis, 56, 143, 147
Derkach, Andrii, 190–191
Dewey, John, 157
Digital devices, discarded, 85
Digital dumpster, 81–85
Digital hoarding, 82
Disinformation as a service, 5, 196–197
Dixie Cups, 75, 94
Domestic masspersonal social engineering, 193–197
Draper, John "Cap'n Crunch," 54, 58
Dreeke, Robin, 103, 148

Drew Theological Seminary, 32
Dual Core, 89–90
Dumpsters, 78–81. *See also* Trashing

Earp, Edwin, 32, 39, 40
Easter Parade of 1929, 95–96, 136
Education, 157–158, 159–160
Ekman, Paul, 148
Elitism, 28–29, 39, 42–44, 62
Enemies of the People list, 192–193
Engineering, in early twentieth century, 29–31
Engineering mindset, 30
Engineering of consent, 11–13, 19, 22, 39–40, 45–46, 63, 71, 75, 152, 157–158
Engineering of Consent, The (Bernays), 38
Epistemic asymmetry, 61, 62
Era-Media TOV, 190
Ethical hacker social engineering, 208–211
Ewen, Stuart, 27–28, 45, 46, 62, 133
Exploding the Phone (Lapsley), 54–55

Facebook
  Cambridge Analytica and, 171–172, 175–176
  dark ads and, 181
  digital traces and, 83
  Internet Research Agency and, 178–180
  manipulative communication on, 1
  as masspersonal communication, 17
  masspersonal social engineering and, 167
  microtargeting and, 14
  political advertising on, 216
  Rally Forge and, 193–194
  Russian election interference and, 2
Facts, love of, 39
Faerie Queene, The (Spenser), 58
Failures, importance of, 206–207
Fake news (term), 5–6
Fallis, Don, 119–120, 122, 125

Fancy Bear, 170
Fischer, Claude, 59
Fleischman, Doris
   American Tobacco and, 95–97, 135–136, 154
   criticism of, 45
   Dixie Cups and, 75
   engineering of consent and, 11–12, 19, 39–42, 45–46, 63
   field of public relations and, 37–38
   in-home observation and, 71
   pretexts and, 22, 93–97
   stereotypes and, 112, 113
Florman, Samuel C., 29–30, 42, 46
Flying Penguin, 124–125, 126
4chan, 2–3
Fox News, 205
Frankfurt, Harry, 118–119, 122, 130, 132, 137, 179
Friedman, Milton, 53
Friendliness, 125–129, 131
Frogs into Princes (Bandler and Grinder), 52
From Counterculture to Cyberculture (Turner), 51
Fugitive Game, The (Littman), 103, 107–108, 143

Game On (talk show), 144
"Game Time!" 208–209
Ghost in the Wires (Mitnick), 98, 103, 108
Gilbreth, Frank, 35
Gilbreth, Lillian, 35–36
Gin, etymology based on, 58
Giuliani, Rudy, 191
Global Sciences Research, 175
Goddard Space Flight Center, 140
Goldstein, Emmanuel, 62
Google Docs, 83
Google Dorks, 84, 86
Grassley, Chuck, 191
"Great Meme War," 2–3

Great Society, 43
Grenfell, Helen, 133
Grinder, John, 52
Group mind, 39–40

Hackers/hacker social engineering
   bullshitting and, 130–132
   intellectual roots of, 50–53
   as interpersonal communication, 59–60
   interpersonal communication and, 59–60
   introduction to, 49–50
   masculinity and, 141–144
   phone phreaks and, 55–56
Hacking (term), 6–7
Hadnagy, Chris
   on control, 144–145
   as ethical social engineer, 208, 211
   on pentesting, 141
   on pretexting, 91, 103
   professionalization of field and, 80–81, 131, 148
   on social engineering, 61, 89, 113
   training offered by, 146
   on uses of social engineering, 159–160
Harbinger, Jordan, 54, 143–144, 145
Hatfield, Joseph, 19, 50, 61, 62, 142
Hawley, Josh, 2
HBGary, 198
Headley, Susan "Thunder"
   at DEF CON, 150
   as phone phreak, 54, 55
   Roscoe Gang and, 143
   transformation of into professional, 147–148
   transition of to hacker, 56
   trashing and, 70, 80, 83–84
   on using sex to manipulate, 145
Heart of Texas Facebook group, 178–179
Hidden Persuaders, The (Packard), 75
Hill, Fiona, 173
Hird, Myra, 77, 79, 82

Hitchhikers, 121, 123, 217
Hoffman, John, 69, 80
Hogan, Mél, 81–82, 215
HOPE conference, 62, 126–128
House, Edward M., 198
Hyper-individualism, 50

Identity play, 98–101
Immigration
　concerns regarding, 31, 32–33
　Taylorists and, 35
Impartiality, 42
Individualism, rise of, 52–53
Individual therapeutic culture, 50, 51–53
Industrialized disinformation, 197
Infoglut, 87
Information warfare, 198
Insikt Group, 196
Instagram, 1, 2, 167
Institute for Propaganda Analysis, 157
Int80, 90
Internet Research Agency (IRA). *See also* Russia
　bullshitting and, 177–180
　methods of, 2–3, 11, 14, 167
　penetrating and, 185–187
　performance metrics at, 185
　pretexting and, 172–174
　Russian sponsorship of, 7
　trashing and, 170–171
Interpersonal communication, hacker social engineering as, 59–60
Interpersonal social engineering, 49–65
Iran, 191–193, 198

Jamieson, Kathleen Hall, 7, 202
January 6, 2021, riot, 2
Jefferson, Thomas, 151–152
Jeffries, Ross, 143, 220
Johnson, Lyndon B., 43

Johnson, Ron, 191
"Joker," 99–101

Kahneman, Daniel, 148
Kaiser, Brittany, 12–13, 172, 175–176, 187, 210, 223
Kant, Immanuel, 76
Karpf, David, 200, 202
Kennedy, Dave, 85
Kitchen Efficient, 35
Knowledge production, 76–78
Kogan, Aleksandr, 171, 175

Labor unrest, attempts to eliminate, 34–35
Landfills, 73, 77, 82
Langer, Ellen, 148
Lankford, Jeff, 108, 109
Lapsley, Phil, 54–55
Le Bon, Gustave, 31, 38
Lee, Gerald Stanley, 27, 31, 38–39, 40
Lee, Ivy
　bullshitting and, 132–135
　on consent engineering, 158
　criticism of, 45
　on crowd and leadership, 157
　"Declaration of Principles" and, 115, 134, 135
　field of public relations and, 37–38, 137
　in-home observation and, 71
　metrification and, 154–155
　news engineers and, 39
　pretexts and, 22
　social engineering process and, 41
Lee, John "Corrupt," 54
Lie machines (term), 5–6
Lingo, use of, 129–130
Lippmann, Walter, 31, 41, 112, 157
Littman, Jonathan, 103, 107–108, 143
Lockheed Martin, 221

Long, Johnny, 81, 88, 131, 145, 148
Lorde, Audre, 223
Ludlow Massacre, 132–135

Malengin, 58
Management theorists, 30–31
Marshall, Jennifer Jane, 94
Masculinity, hacker social engineering and, 141–144
Masses/crowd, 31, 38–40, 151–152, 166
Mass media, emergence of, 37
Mass messaging, 200
Masspersonal communication, 15–18
Masspersonal social engineering
  ameliorating, 199–225
  case studies of, 169–188
  definition of, 4
  domestic, 193–197
  ethics of, 208–211
  heuristic for, 168–169
  introduction to, 1–23, 165–168
  Iran and, 191–193
  potential effectiveness of, 201–208
  potential responses to, 211–225
  Russia and, 189–191
  since 2016, 188–197
  social engineering process and, 168–169
  summary of, 197–198
Mass social engineering
  bullshitting and, 132–137
  definition of, 61
  rise and fall of, 27–47
Masters of Deception (hacker group), 69–71
Masters of Deception (Slatalla and Quittner), 69
Mears, Daniel, 120, 121
Mechanical Turk, 175
"Me Decade," 50
Mediacracy: American Parties and Politics in the Communication Age (Phillips), 43–44

Media Education Foundation, 217
Media literacy, 217–219
Media Manipulation Casebook, 217
Media penetration, 153–155
Mentor, The, 58
Mercer, Robert, 171, 176
Metrification, 153–155
Microtargeting, 14, 167, 187, 196
Mills, C. Wright, 44–45
*Mindf*ck* (Wylie), 11
Mitnick, Kevin
  as ethical social engineer, 208
  on human as weakest link, 59
  interpersonal con artistry of, 19
  Motorola attack by, 204
  on people skills, 131
  as phone phreak, 54
  on Podesta phishing attack, 10
  reputation of, 90–91, 97–98
  Roscoe Gang and, 143
  social capital and, 101–110
  social engineering and, 9, 115, 159
  as social joker, 100–101
  stereotypes and, 112–113
  transformation of into professional, 147–148
  transition of to hacker, 56
Mitnick Security, 147
Monitor, USS, 206
Moore, Roy, 194–195
Motorola, 204
Mueller Report, 173–174, 178
Mukerji, Chandra, 121, 122, 123, 128, 130, 217
Museum of Public Relations, 136
Music Walrus, 175–176

Nabu Leaks, 190
Narratives, 222–223
Narrative switching, 180
Narrowcasting, 17
NASA, 140
National Review, 43

National Soap Sculpture Committee, 94–95
NEC, 108–110
Neoliberal economics, 50
Neoliberalism, 53
Nerd masculinity, 142–143
Networked Propaganda, 6
Neuro-linguistic programming (NLP), 52, 143, 144–145
New Age Politics (Satin), 52
New Communalist thinking, 50–51
News engineers, 38–39
Newspapers, 151–152
Nix, Alexander, 3, 176–177, 183–185, 187–188, 209–210
No Tech Hacking (Long), 81, 131, 148

Ocasio-Cortez, Alexandria, 2
On Bullshit (Frankfurt), 118, 132
"One best way," 34
Online Marketing Rockstars, 176, 188
Only News, 190
Open-Source Intelligence (OSINT), 72, 86–88
Operation Igloo White, 207
Organizational structures, 103–106
Osborne, Matt, 194–195
O'Sullivan, Patrick, 16–17, 64, 167–168
Oxford Internet Institute, 196–197

Pacific Bell, 104–106, 147
Packard, Vance, 74–75, 82
PACs, 216
"Paper bullet" metaphor, 152
Parasite, The (Serres), 99
Passing, 98–99, 100, 109
Pelosi, Nancy, 2
Penetration/penetrating
  bullet metaphor and, 151–153
  Cambridge Analytica and, 187–188
  as goal, 22
  introduction to, 139–141
  media and, 153–155
  physical, 60
  potential responses to, 219–225
  professional, 146–150
  Russia and, 185–187
  sexual conquest metaphor for, 141–146
  uses of, 155–161
Penetration testing (pentesting), 139–140, 146–150, 208, 211
Persona management, 198
"Perspecticide," 182
Pfeffer, Carla, 98–99
Phillips, Kevin, 43–44
Phishing, 9–10
Phone phreaks, 49–50, 53–60, 115, 117–118, 123–130
Phunware, 5, 7, 195–196, 198, 207, 223–224
Physical penetration of organizations, 60
Pickup, The (podcast), 144
Pickup artists, 142–146
Pittsburgh Study, 33
Planned obsolescence, 74–75, 85
Plausibility, 123
Podesta, John, 9–10, 167, 169–170
Political communication, 46–47
Porat, Aelon, 208–209, 211
Presidential election of 2016. *See also* Cambridge Analytica; Clinton, Hillary; Internet Research Agency (IRA); Trump, Donald
  bullshitting and, 177–185
  methods used for, 166–168
  overview of interference in, 2–3
  penetrating and, 185–188
  Podesta phishing attack and, 9–10
  pretexting and, 172–177
  trashing and, 169–172
Pretexts/pretexting
  authenticity and, 92–93
  Cambridge Analytica and, 174–177
  hackers and, 91–93
  identity play and, 98–101

Pretexts/pretexting (cont.)
  introduction to, 21–22, 89–91
  mass social engineering and, 93–96
  phone phreaks and, 124–125
  potential responses to, 215–217
  Russia and, 172–174
  social support structures and, 103
  stereotypes and, 110–113
  structural factors for, 101–110
  tiring nature of, 103
Privacy guides, 212
Proctor & Gamble, 94–95
Project Birmingham, 195
Propaganda (term), 6
Proud Boys, 192–193
Psychographics, 187, 200
Psychological obsolescence, 74–75
Psychological warfare, 158–159
Publicity (Lee), 134
Public Opinion (Lippmann), 41
Public Opinion Quarterly, 155
Public relations
  emergence of, 36
  mass social engineering and, 37–42
  rejection of social engineering by, 45–46
  role of, 28
Public Relations Society of America Volunteer Chapter, 216
Putin, Vladimir, 209

Quittner, Joshua, 69

Racial stereotyping, 106–110
Radcliffe, Jenny, 145, 208, 211
Rake, Derek, 144
Rally Forge, 5, 193–194, 195, 207
Rathke, William, 69
Recognition, 99, 101, 111, 112–113
Recorded Future, 196
Red Boxes, 56
Richards, Ellen Swallow, 73
Rid, Thomas, 10–11, 199, 203

Right-wing media ecosystem, 205
Rivera, Geraldo, 70, 80, 147
Rockefeller, John D., Jr., 132–135, 154–155
Roscoe Gang, 143, 147
RT, 2, 12
Russia
  bullshitting and, 177–180
  continued social engineering efforts by, 10–11
  election interference from, 2–3, 4–5, 7, 22, 165, 166–168, 189–191, 202
  mass propaganda efforts by, 12
  penetrating and, 185–187
  phishing by, 9–10
  recent masspersonal social engineering and, 189–191
  trashing by, 169–171

Sage, Robin, 145
SAGE air defense system, 207
Sanitary engineers, 73–74
Satin, Mark, 52–53
Scanlan, John, 76–77
Schneier, Bruce, 221
Schweickert, Molly, 181, 188, 213
Science of coercion, 220, 221
Scientific Managers, 33–36, 39
SCL Group, 177
Secret History of Hacking, The, 57
Seduction community, 143
Self-curation, 82–83
Self-help culture, 50, 51–53
Self-regulation, 216
Sensual Access (De Payne), 143
Sentiment analysis, 154–155
Serres, Michael, 99
Server farms, 215
Settlement House Movement, 32–33
Sex Compass, 175
Sexual conquest metaphor, 141–146, 188
Shay's Rebellion, 151
Shimomura, Tsutomu, 106–107, 109

"Shogun Method," 144
Simpson, Christopher, 153, 155, 220, 221
60 Minutes, 147
Skeptic TOV, 190
Slatalla, Michelle, 69
Smith, Adam, 53
Soap carving competitions, 94–95
Socialbots, 173–174
Social capital, 101–110
Social control, 32–33
Social Engineer, The (Earp), 32
Social engineering. *See also* Hackers/hacker social engineering; Masspersonal social engineering
 current versions of, 46–47
 in early twentieth century, 19–20, 30–31
 interpersonal, 49–65
 interpersonal communication and, 59–60
 introduction to, 8–15
 managerialist, 33–36
 meanings of, 166
 as pejorative, 42–46
 phreaking and hacking via, 57–59
 relating mass and interpersonal, 60–63
 scaling up of, 22
 social control through, 32–33
 as term, 8
Social Engineering in Cincinnati, 33
Social Engineering in IT Security (Conheady), 63, 81, 91, 103, 145, 149, 159
Social Engineering: The Art of Human Hacking (Hadnagy), 80–81
Social Engineering: The Science of Human Hacking (Hadnagy), 141, 159
Social-Engineer.org (podcast), 89, 143–144, 148
Social Gospel Christian activists, 30, 33
Social media. *See also individual platforms*
 as masspersonal communication, 17–18
 self-regulation and, 216

Social stereotyping, 110–113
Social support structures, 103
Sociological Imagination, The (Mills), 44
Sociologists, 30
SolarWinds incident, 219
Soundscapes, 92
Speed seduction, 143
Spenser, Edmund, 58
Sprint, 126–128
Sputnik, 2, 12
Sreenivasan, Rahul, 170
Stanford Internet Observatory, 193–194
State of Black America Report (2019), 173
Stealth marketing, 210
Stereotypes
 bullshitting and, 119
 racial, 106–110
 reductive, 91
 social, 110–113
Stokke, Andreas, 119–120, 122, 125
Strategic communication, 46, 198, 220
Street, Jayson, 139–140, 145, 150

Tailored spam, 17
Tandem stacking, 117
TAP (Technological Assistance Program), 56, 70, 79–80, 124, 125, 126, 128
Tarbell, Ida, 133
Taylor, Frederick Winslow, 34–35, 39
Technocratic dominance, 61, 62, 142
Technology and Social Change Research Project, 217
Teleological replacement, 61
Telephone Electronics Line, 56, 70
Teresi, Denny, 54
Terry, Dennis, 57
Three-part relations and communication, 99
Torches of Freedom march, 96, 135–136
Trans identities, 98–99
Transit fraud, 101–102

Transmission view/model of communication, 153, 221
Transnational capitalism, 103–106
Trash, collecting information from, 21–22
Trashing
 digital, 81–85
 guides to, 78–81
 Internet Research Agency (IRA) and, 170–171
 introduction to, 69–72
 OSINT and, 86–88
 philosophy of garbage and, 76–78
 Phunware and, 195–196
 potential responses to, 212–215
 Russia and, 169–171
 social engineering as possible through, 72–73
 societal practices and, 73–76
True/false dichotomy, 5–6
Trump, Donald. *See also* Presidential election of 2016
 Cambridge Analytica and, 13, 175, 176, 183
 campaign slogans of, 181
 Capitol insurrection and, 2
 efforts to help, 202–203
 election of, 2–3
 IRA influence on campaign of, 186
 microtargeting and, 14
 Phunware app for, 223–224
 Russian election interference and, 190–191
"Trust Me" (song), 89–90
Tullock, Gordon, 53
Turnbull, Mark, 177, 181, 187
Turner, Fred, 51
Turning Point Action, 194
2600, 56, 62, 70, 79, 84, 124
20/20, 70, 147
Twitter, 1, 2, 17, 83, 167, 173–174
Tyrrell, Emmett, 44

Unseen Power, The (Cutlip), 43
US Commission on Industrial Relations, 134
US Cyber Command, 189
US Department of Justice, 185–186
US Office of Naval Intelligence, 174
US Office of the Director of National Intelligence (ODNI), 12, 14, 189
US Senate Select Committee on Intelligence, 186

Verizon Data Breach Investigations Report, 59
"Vishing," 131

Waste Makers, The (Packard), 74
Waste management, 77
Wasting, infrastructure for, 72–74
Wealth of Nations, The (Smith), 53
Whole Earth Catalog (Brand), 51
Whole Earth 'Lectronic Link (WELL), 51
Wiener, Norbert, 207
WikiLeaks, 2, 10
Williams, Raymond, 221, 222
Wired, 51, 85
Wolfe, Tom, 50
Won, Joseph, 106–107
World War I, 33, 36
Wozniak, Steve, 55
Wylie, Christopher, 11, 165, 172, 181–185, 210, 220, 223

Yang, Gordon, 221–222, 223
"Yellowface minstrelsy," 106–110
Yonder (formerly New Knowledge), 194
#YourSlipIsShowing hashtag movement, 217–218